Interactivity

Interactivity

New media, politics and society

Alec Charles

Peter Lang Oxford

Peter Lang Ltd
International Academic Publishers
52 St Giles, Oxford, OX1 3LU
United Kingdom

www.peterlang.com

A catalogue record for this book is available from the British Library

ISBN 978-1-906165-44-4

Printed in the United Kingdom
by TJ International

Contents

Acknowledgements

Grateful thanks for their support are due to my friends and colleagues James Crabbe, Peter Dean, Ivor Gaber, Kelly Hallam, Michael Higgins, Luke Hockley, Dan Jackson, Malcolm Keech, Mary Malcolm, Carsten Maple, James Morrison, Bill Rammell, Heather Savigny, Jon Silverman, Mick Temple, Liesbet van Zoonen, Alexis Weedon, Garry Whannel and Dominic Wring. Thanks are also due to my students, whose feedback and suggestions have been invaluable, and to all those who kindly gave their time to contribute their comments to this study: Jaak Aab, Andres Aarma, Danah Boyd, Moira Burke, Iain Dale, Gonzalo Frasca, Jane Griffiths, Aleksei Gunter, Joe Hewitt, Andrew Keen, Adam Kramer, Davin Lengyel, Tim Loughton, Tarvi Martens, Austin Mitchell, Mark Oaten, Mart Parve, Larry Sanger, Tex Vertmann and Linnar Viik. And particular thanks are of course due to Lucy Melville, Mette Bundgaard, Holly Catling and Mary Critchley at Peter Lang.

CHAPTER I

Strange New World

New media technologies have not changed the world. The sun has not fallen from the sky. The rivers still run to the sea. The poor remain with us, and many, many millions continue to toil on the land and to hunger in their beds. We still, for the while, inhabit these vile, frail, mortal shells.

Yet it seems clear that these technologies have radically affected very many lives. High street stores are closing down, our smartphones oblige us to take our work wherever we go, and nobody writes letters anymore. Bill Gates has killed off the typing pool (although, one should stress, not literally so). As the likes of Keen (2008), Papacharissi (2010) and Gregg (2011) have observed, our daily lives and labours have altered in ways unimagined a decade or so ago, and western societies have been revolutionized (although, again, not literally so).

But if we are not in Kansas anymore, then we are also still a very long way from Oz.

So here we are, caught between two worlds, dynamically suspended in a state of radical transition and of fundamental uncertainty.

Which not wholly unfamiliar situation might lead one not entirely unfairly to ask: So then – *what's new?*

The Knock of the Shoe

According to Hodson and Sullivan (2012: 71), history's original saboteurs were reputedly those fifteenth-century Dutch textile workers so-called because they would throw their *sabots* (wooden shoes) into the machinery in order to make it clog up and break down. (But this is not why clogs are called clogs.)

This book is not, however, intended as an act of Luddite or techno-phobic or ever cyberpessimistic sabotage. There does not appear to be anything inherently wrong with the new media technologies which have come to dominate post-industrialized civilization in the early years of the twenty-first century; nor indeed does there seem for the most part to be anything essentially, overwhelmingly or unprecedentedly meretricious in the uses to which these technologies are put. This book will nevertheless call into question – repeatedly and, one fears, with an increasingly degree of foot-tapping scepticism – the cure-all claims with which many in the technology industry, in the media and in academia have attempted to brand these tools. It will ask whether the modes of interactivity promised by contemporary media forms (from voting on *Big Brother* to friending on Facebook) in fact promote any useful modes of political and social interaction among their audiences and users.

One is tempted at this point to invoke Eric Partridge's gloss for the derivation of *sabotage* – from the French *saboter*, 'to knock with the foot, hence to shake or torment' (Partridge 1991: 578). This book, in these terms, at once recognizes the cultural shock provoked by new media technologies and forms – and attempts to shake up (although not perhaps to torment) some of the responses and assumptions that have accompanied this much-vaunted revolution. In the context of an earlier period of technological revolution, the art critic Robert Hughes (1991: 15) once wrote that 'the speed at which culture reinvented itself through technology in the last quarter of the nineteenth century and the first decades of the twentieth, seems almost preternatural' – and there seems something similarly disturb-ing in the upsetting of naturalized paradigms that characterizes the early twenty-first century's shock of the new.

Palfrey and Gasser (2008: 2) have written of a new generation of 'digi-tal natives' – a generation which interacts in ways strange to those outside this culture – a generation which, they say, will eventually transform global politics (Palfrey and Gasser 2008: 7). This book examines the start of this process of transformation.

This book is, as the title suggests, about interactivity – or rather it is about the ways in which the declarations or illusions of heightened interactivity advanced by new media technologies and applications offer users a deceptive sense of their own active participation in political and social processes, and thereby undermine users' desires and potentials for actual, meaningful and impactful engagement in such processes. This book is about that paradox – the way in which a self-avowedly democratizing agglomeration of media forms might effectively counteract and reverse processes of democratization – and therein about the absurdity of that paradox. We may love these new technologies and their uses – *like we don't all use them – like we don't all depend on them* – without that love being irretrievably unconditional; we may be enthusiastic without being unquestioningly fanatical; we may rely upon them without putting our absolute trust and faith in them, in what they are in themselves and excluding all else. There is clearly nothing fundamentally wrong with new media, communication and information technologies: technology in itself is neither nemesis nor panacea; it is at best merely a facilitator. The problems arise when we come to see it as a solution in itself. That is what this book is about.

Reductio ad Absurdum

This book is also therefore about the absurdity – the definitively paradoxical nature – of contemporary existence, existence, that is, in the information society of the post-industrial world, a domain of irreconcilably clashing paradigms: it is about what happens when, for example, as Jesse Rice (2009: 21) puts it, 'our brilliant but imperfect humanity collides headlong with Facebook's brilliant but imperfect technology.' The twentieth-century French philosopher Albert Camus (1975: 13) identified the feeling of absurdity as resulting from a fundamental existential disconnection, suggesting that a 'divorce between man and his life [...] is properly the feeling of absurdity.'

This then is the extent of the absurdity we face, an absurdity mod-
elled and thereby aggravated by the forms and genres of contemporary
mass media culture. The virtual society – this civilization of electronic
government, digital play, reality television and online networking – gen-
erates a world in which the empty fantasies of *World of Warcraft* parallel
the prevailing condition of absurdity experienced by a civilization which
has sought to wage a War on Terror. This book explores how these media
forms perpetuate and mirror this absurdity, and suggests that, although we
may claim not to have started the fire, it appears we have been attempting
to put it out with gasoline.

The absurdist short-story writer Jorge Luis Borges (1970: 42) famously
imagined the fantasy world of Tlön, a world conceived as a theoretical
model by intellectuals, a virtual world whose structures come to overwhelm
the real world: 'How could one do other than to submit to Tlön, to the
minute and vast evidence of an orderly planet? It is useless to answer that
reality is also orderly. Perhaps it is, but in accordance with divine laws – I
translate: inhuman laws – which we never quite grasp. Tlön is surely a
labyrinth, but it is a labyrinth devised by men, a labyrinth destined to be
deciphered by men.' We have invented a new world to replace the incom-
prehensible cosmos created by a divine being, and that world has become
the reality which we now inhabit. Yet this artificial universe, though absolute and incontrovertible, remains unintelligible to us. This is perhaps the
continuing absurdity of the human condition – a double absurdity in
that the paradigm developed to overcome our existential absurdity has
proven even more paradoxical, even more absurd, than the obsolescent
theocentric perspective it was designed to supersede. In order to bridge
Camus's divorce between subjectivity and existence, we have constructed
a new, anthropocentrically bespoke and irresistible mode of being – one
from which we, nevertheless, feel even more alienated. There yet remain
defining and perhaps unerasable traces of our abandoned paradigms, and
the inconsistencies and contradictions between these and the perspectives
which we are now adopting propagate a sense of inevitable and almost
inconsolable absurdity.

Almost inconsolable – this absurdity is only *almost* inconsolable insofar as, as Camus (1975: 53) proposed, the solution to this dilemma is to recognize the inevitability of that absurdity and to accept the impossibility of transcendental meaning: 'life [...] will be lived all the better if it has no meaning.' To become reconciled to the irreconcilability of existence is, for Camus (1975: 63), to embrace 'the absurd world [...] in all its splendour and diversity.' And yet, because this is clearly nearly impossible, we appear to return time and again to all of those artifices (e-democracy, video games, the social networking site, reality TV) which pledge to banish the absurdity and to install each of us in our rightful place as the controlling and beloved agent, literally the *avatar*, the divinity incarnate, of our existence, and which, in doing so, only serve to exacerbate the extent of our existential alienation.

Youtopia

Zizi Papacharissi (2010: 3) has suggested that new communicational technologies tend to provoke 'narratives of emancipation' which are framed between 'utopian and dystopian polarities.' The cyberutopianists, cyberenthusiasts or cyberoptimists, as they are sometimes called, tend to suppose that new media, information and communications technologies offer a panacea for all of the ills of the world and thus herald an electronic utopia of democratic, economic and educational freedoms and rights on a global scale – advancing a 'technotopia' (Storsul and Stuedahl 2007: 10) or 'a New Alexandria' (Koskinen 2007: 117). As Tiffin and Rajasingham (2003: 26) have suggested, the Internet has been seen by some as 'a democratic virtual meeting place [...] a new Museion of Alexandria.' Indeed in the earlier days of mass Internet use Tiffin and Rajasingham (1995: 7) had gone rather further in their enthusiasm for the potential of this technology: 'VR offers us the possibility of a class meeting in the Amazon Forest or on top of Mount Everest; it could allow us to expand our viewpoint

to see the solar system operating like a game of marbles in front of us, or
to shrink it so that we can walk through an atomic structure as though it
was a sculpture in a park; we could enter a fictional virtual reality in the
persona of a character in a play, or a non-fictional virtual reality to accompany a surgeon in an exploration at the micro-level of the human body.'
One is tempted to recall in this context the avowed cyberrealist Andrew
Keen's scepticism as to the 'mad utopian faith in our ability to conquer the
physical world through virtual reality' (Keen 2008: xviii).

Some have lauded new media technologies as the key to the global
establishment of liberal democracy, often without recognizing the limitations of these tools, their potential for appropriation by reactionary and
totalitarian institutions of power or for that matter the questionable universal appropriateness and value of such a political system. J. Hunter Price
(2010: 1) has, for example, pointed out that, as these technologies have
been increasingly deployed by opposition campaigners in dictatorial states,
western commentators have grown 'even more confident in their conviction
that new media will lead to a wave of democratization.' This is the dream
of the cyberoptimists. As Joss Hands (2011) has declared, @ *is for Activism*.

The cyberpessimists, by contrast, tend to suppose that George
Orwell was right: that we are increasingly living in a surveillance society,
a hegemonized and homogenized realm, a media dictatorship of totalizing
ideologies and corporate hyperpower – the world which Andrew Keen
(2008: 166) has dubbed '1984, version 2.0.' This is a mode of Orwellian
dystopia which is as consensual as it is totalitarian: as Keen (2008: 175)
reminds us, 'the age of surveillance is not just being imposed from above
by aggregators of data. It's also being driven from below by our own self-
broadcasting obsession.' It is the YouTube youtopia.

This is a world in which new media technologies sponsor fanaticism
and terrorism as much as popular liberation. This is a world in which the
notion that the Internet empowers the powerless is, as Morozov (2011a: xii)
suggests, no more than so much 'cyberutopianism'. The cyberpessimistic
position tends to be epitomized by anxieties as to what Talbot (2007: 173)
has identified as the 'increasing corporate control of the Internet.' Such
reports as that of the 2010 agreement between search engine Google and
Verizon (America's largest telecommunications company), described by

The Guardian newspaper (5 August 2010) as 'a deal that could bring an end to "net neutrality"', have led to fears of a double speed Internet that would reinforce digital divide (Frau-Meigs 2007: 48) and undermine the democratic potential for which (in everything from electronic government to the social networking site) the Internet has been celebrated by the cyberoptimists.

Google again came under fire in March 2012 when the European Union's Justice Commissioner Viviane Reding declared that changes to Google's privacy policy – changes which allowed the search engine to share user data with its other platforms (including Gmail and YouTube) – breached EU law. Her announcement not only prompted the *BBC News* website to offer its readers instructions as to how to delete their Google histories, but also echoed and amplified concerns that have increasingly dogged the corporate giants of the Internet industry.

The problem, stated simply, is this: in the world of the Web there is no such thing as a free search (or a free social network or host for your emails, videos or blogs, for that matter). Organizations like Google and Facebook rely on advertising in order to generate revenues essential not only to sustain their burgeoning corporate and technological infrastructures but also to reward their shareholders. In May 2012 Facebook launched its first stock market flotation, offering $18 billion dollars' worth of shares – the result of the highest valuation of Internet stock in an initial public offering since Google had raised $1.67 billion dollars in 2004, but still only a fraction of the estimated $104 billion valuation of Mark Zuckerberg's company. As the users of these services remain unwilling to pay for them themselves, these organizations have therefore looked to advertisers to meet the costs not only of development but also of dividends. Yet what advertisers want from these companies is targeted access to their hundreds of millions of customers, based upon data harvested as to these customers' online preferences and personal information.

This is why one's Facebook page features adverts for products for which one has declared a fondness in one's list of likes. This is why some sites deny access to computers set to refuse the temptations of those innocently named 'cookies' – packets of data which track and hold information on user preferences. (Indeed research published by the online privacy solutions organization Truste in April 2012 suggests that users encounter an average of 14 trackers per page on Britain most popular websites and that 68 per cent of these trackers belong to third-party organizations. In Europe such has been the concern as to the intrusiveness of cookies upon users' privacy that new laws came into force in May 2012 obliging websites to seek users' consent before harvesting data via cookies – although in the UK the Information Commissioner's Office warned of widespread non-compliance with the terms of this legislation.)

This is also why, conversely, the Internet's fifth most popular website, the not-for-profit Wikipedia, regularly solicits donations from its users in order to maintain its modest contingent of staff and servers while avoiding the need to court the advertising market. And this is therefore the reason why in March 2012 – in response to the Google privacy controversy – the director of the civil liberties group Big Brother Watch told the BBC that he feared that Google was now 'putting advertisers' interests before user privacy.'

In August 2010 Google boss Eric Schmidt had informed *The Wall Street Journal* that he did not believe that privacy regulation was necessary – because his users could abandon the site if Google did anything with their personal information that they found 'creepy'. But do Internet users really have much of an alternative? Indeed one might ask whether they might have the right to demand higher ethical standards from these self-proclaimingly idealistic companies – when, as Fuchs (2011) has suggested, the net's most popular platforms allow their users scant opportunities for participation in the development of their own terms of use policies.

Around the turn of the millennium the architects of Web 2.0 (the interactive incarnation of the Internet) had heralded the online age in democratic, communitarian and egalitarian terms which promised the technotopia of a global village. So where did it all go wrong? Today's moguls of the net have this much in common with the leaders of any revolutionary movement: they have discovered that the maintenance of power, once achieved, all too easily seeks recourse to the established hierarchies, structures and interests which their idealism had once sought to overthrow.

Yet the optimism of many cyberenthusiasts continues apparently una-bated. The optimists or cyberenthusiasts represent a position summed up by Bolter and Grusin (2000: 59–60) in the following terms:

> They tell us, for example, that when broadcast television becomes interactive digital television, it will motivate and liberate viewers as never before [...] that hypertext brings interactivity to the novel [...] that the World Wide Web [...] can reform democracy by lending immediacy to the process of making decisions.

This position is neatly challenged by Koskinen (2007: 125):

> What is the point in selling the idea that digital TV makes it easier for us to order pizza when any modern city already provides plenty of opportunities for ordering pizza? [...] Take the notion of interactive narratives [...] No one in his right mind can write an alternative ending to the story of Jesus Christ. Or what is the point in taking *Romeo and Juliet* and attempting to 'improve' its dialogue by making it interactive?

As Liesbet van Zoonen (2010) has suggested, although there has been a lot of utopian talk about what Web 2.0 can do, its failure to achieve those socio-political ambitions has been quite embarrassing. Papacharissi (2010: 20) points out that processes of commercialization have simply turned new media into versions of older media – 'expanding shopping catalogs for the consumer, but not affording democratic options for the citizen.' The Internet might then from this perspective come to seem no more than a domain of cynical commercialism, of user narcissism and paranoia, of pornography, price comparison websites and home movies of kittens and *Jackass*-style adolescent stunts.

However, despite the apparently irreconcilable nature of the cyber-optimistic and cyberpessimistic stances, we might concur with both the cyberoptimists and the cyberpessimists on one central theme: that the evolution and adoption of new media technologies heralds a paradigm shift in our notions of politics, society and subjectivity. This is not, of course, the end of civilization as we know it, but a transformation of our model of civilization. This transitional period defies rationalization in traditional terms, generating instead paradoxes whose absurdity reveals an impasse within received notions of identity, meaning and societal progression.

This book does not suggest that this is a bad thing *per se*, but proposes that our time-honoured theories and ideals of individual self-determination, social interaction, representative democracy, military conflict, factual representation and material actuality are becoming increasingly irrelevant to the conditions we are heading towards: that these perspectives are coming to represent what Papacharissi (2010: 8) has called 'metaphors that no longer work.' This is not the critical apocalypse of nuclear Armageddon, of fundamentalist terrorism or of anarchic revolution; this seems an ongoing repositioning of history itself, the propagation of a world without material history, a virtual gameworld without the immediacy of an awareness of material responsibility. Like global warming, its future extent is (thus far) almost undetectable to the naked eye. The transfiguration of the western cultural paradigm takes place virtually unnoticed, save for that mild feeling of discomfort – of metaphysical alienation or absurdity – of which some complain. And so the frog boils. This is how T.S. Eliot (1925) famously said the world would end – *not with a bang but a whimper.*

The World According to Google

A graphically convenient way to demonstrate the ongoing paradigm shift from the traditional material-historical perspective towards a homogeneous, mass-mediated, globalized world view is to conduct a Google image search on the word *Homer*. The overwhelming majority of the images generated by the search will show Bart Simpson's father, rather than the legendary Greek poet and originator of western literary civilization. This prioritization of immediately contemporary, homogeneously popular culture is evidenced by a variety of more formal studies. In August 2006 the BBC reported that a survey commissioned by an online game show devoted to modern popular culture had shown that more Americans knew who Harry Potter was than Tony Blair. While 57 per cent of the 1,213 participants surveyed recognized Harry Potter, less than half recognized the UK's then

Prime Minister. The survey also showed that, while six out of ten people knew Homer Simpson's son was called Bart, only a fifth could name one of Homer's ancient Greek epics.

A sense of history (in its traditional sense) is diminishing. In December 2008 it was reported that, in the year of his election to the U.S. Presidency, Barack Obama trailed in third place in the league of Yahoo's greatest number of searches behind pop singer Britney Spears and World Wrestling Entertainment. In June 2011, it was reported that Obama had gained third place in Twitter's most followed tweeters: this time beaten by teen pop sensation Justin Bieber and surreal pop sensation Lady Gaga. Political history is runner-up to populist mass media culture. In November 2009 *The Daily Telegraph* informed its readers that a survey had discovered that 5 per cent of British schoolchildren believed that Adolf Hitler had been the coach of the German football team – while one in six youngsters believed that Auschwitz was a Second World War theme park and one in 20 thought the Holocaust had been a party to celebrate the end of the war.

A report published in August 2010 by Britain's Office of Communications announced that the British were spending almost half of their waking hours watching television or using mobile phones and other communication devices. The report added that the British were also immersed in several types of media at the same time – on average fitting nearly nine hours of daily media engagement into just over seven hours. Research into habitual Internet use has demonstrated that online activities are altering the very structures of our brains: Kanai et al. (2011) have, for example, suggested a direct correlation between the number of friends held on social networking sites and the density of brain matter in the right superior temporal sulcus, left middle temporal gyrus and entorhinal cortex.

Andrew Keen (2008: 160) has noted a study by Stanford University in which one in eight adults were shown to manifest symptoms of Internet addiction. As Gary Small, a researcher into this field at UCLA, told *The Guardian* newspaper in August 2010, 'the Internet lures us. Our brains become addicted to it. And we have to be aware of that, and not let it control us.' In April 2011 a report in *The Daily Telegraph* cited research conducted by the University of Maryland's International Center for Media and the Public Agenda which found students across the world admitted being so 'addicted' to their mobile phones, laptops and social networking sites that 80 per cent suffered significant physical and emotional distress when stripped of these technologies for just one day. In May 2012 researchers at Norway's University of Bergen published a study which likened forms of Internet dependency to drug and alcohol addiction.

Thus Andrew Keen has argued:

> It's something that many people would struggle to live without. I'd struggle to live without it. If you haven't been on it for a few hours, you get withdrawal symptoms. The challenge is to make it a more regular part of people's lives, rather than something they're on all day. Mobile technology makes it more addictive – you can take the Internet everywhere.

If this represents a form of addiction then it is one which can display particularly unpalatable manifestations. In April 2012, for example, a cross-party report from the UK parliament revealed that 80 per cent of 16-year-olds regularly access porn online – and that a third of 10-year-olds have also done so; or, as the *Daily Mail* put it, a 'generation of children is growing up addicted to hardcore internet pornography.'

The Immaterial World

As electronic media move our perspectives beyond the traditional constraints of space and time, and restructure the ways we think and perceive, the phenomena of existence appear to be becoming increasingly immaterial. This has been going on for quite some time. It was, in fact, about the time that Madonna announced we were living in a material world that we started to realize we were no longer were. That was in 1984, but even George Orwell could not have seen what was coming. The Internet Activities Board had been established just the previous year. The virtual world had begun.

Welcome, therefore, to *the desert of the real*. This sentiment is advanced by that great prophet of the virtual world, the sociologist Jean Baudrillard, in one of his last books (Baudrillard 2005: 27). In deploying this phrase Baudrillard is alluding at once to Slovenian philosopher Slavoj Žižek's book of 2002, *Welcome to the Desert of the Real*, and to that line of Laurence Fishburne's from the 1999 film *The Matrix* – from which Žižek's book takes its title. The Wachowski brothers' film – which features a copy of Baudrillard's earlier work *Simulacra and Simulation* – itself takes that phrase, *the desert of the real*, from the very first page of that book, Baudrillard's own seminal work (Baudrillard 1994: 1). This most postmodern cycle of referentiality offers a pertinent metaphor for the contemporary condition of historical disorientation. A sense of originality and authenticity has been lost within this post-material reality – as is witnessed in this allusion to an allusion to an allusion to one's own original object (Baudrillard's allusion to Žižek's allusion to the Wachowskis' allusion to Baudrillard). The allocation of provenance (and therefore the possibility of historical logic) becomes unfeasible in a cultural environment dominated and epitomized by this recurrence of reflection, this palimpsest of simulacra (copies without originals, shadows without substances, reflections without objects). The desert of the real is not, as Laurence Fishburne's Morpheus supposed, the material world: it is the meaningless void of the virtual realm, the world that most of us in the western world spend so much of our time inhabiting. Yet, although there is no authenticity, we are still tormented by a nostalgic yearning for it. Without authenticity as its opposite, there can be no true fakery; yet our desire for authenticity necessitates that fakery (if we cannot have the real thing, we feel the need to fake it). This is the paradox and the tragedy of the simulacrum.

In *The Gulf War Did Not Take Place* Baudrillard (1995: 85) imagined that the New World Order would be a hypermediated one. By the time of his death in March 2007 it appeared that Baudrillard's prophecy had been fulfilled. The world, as he came to see it, was transcendentally mediated: real-time media had come to define real time, reality television had determined a televisual reality – a reality which was at once perfect, totalizing, virtual, integral, utopian and hyperreal, a post-material ultra-reality (Baudrillard 2005: 27). Baudrillard (2005: 32) witnessed an eventual paradigm shift beyond the simulacrum – 'that which hides the absence of truth' – into a reality which now acknowledges (and indeed takes for granted) truth's absence.

Baudrillard depicts a world without distance or difference, a world beyond history and memory, in which the media can no longer distinguish between existence and representation, a world which has witnessed a blurring of the distinctions between democracy and television, between war and games, between society and virtuality, between fact and fantasy. This is a world in which technology for the augmentation of reality is now available in a pocket-sized device – such as, for example, the Nintendo DSI which (in the words of *The Guardian*'s Charlie Brooker) can 'superimpose pig snouts or googly eyeballs over your friends' visages when you point it at them.'

In June 2010 Bryan O'Neil Hughes, the product manager for Photoshop, told *The New York Times* of the growing trend to augment one's own photographic image: 'Everybody is suddenly representing themselves to the world this way, and you see people doing different things, just to make themselves look better or stand out – anything from mild retouching to putting themselves in pictures and making themselves look like paintings, and all that "Avatar"-y sort of stuff.' This augmented reality often seems more attractive the material world. In January 2010, for example, the *Daily Mail* reported what it called 'the Avatar effect' – an unforeseen response to James Cameron's computer-generated 3-D epic – a level of depression (reaching suicidal proportions) at the cinema-goer's return to the real world. We need look no further than the much-reported case of David Pollard – who in 2008 left his wife (whom he had married in Second Life) for a woman he met in Second Life – and with whom his wife caught him cheating in Second Life – to marry his new partner both in Second Life and in real life – to see that the ontological prioritization of the virtual world is already, for some at least, a *fait accompli*.

The processes of mediation have become so hegemonic and unrestrainedly extreme that it is increasingly difficult to tell what is real and what is not – insofar as the real is defined by those processes of mediation. Reality television, has for instance, become so absurd, so beyond any sense of what might once have been perceived as realistic, that one may be forced to conclude that it only maintains its epithet by virtue of the fact that it is now such television which determines how we measure and recognize reality. Thus, having shifted the centre of normative reality, reality television programme formats and practices have, in their most extreme manifestations and in their attempts to court ratings, become ever more eccentric. In January 2005, for example, it was reported that Fox Television had launched a short-lived reality TV show entitled *Who's The Daddy*, a programme in which adopted children would stand to win $100,000 by guessing the identities of their biological fathers from a line-up of eight men. Two months later German TV channel RTL II announced that its new series of *Big Brother* would feature not just a house but an entire village and would have no end date set. (One has flashes of Patrick McGoohan's 1960s fantasy series *The Prisoner*.) The show was cancelled the following February as a result of poor ratings. In both of these cases we may assume that the failures of these programmes suggest that their creators were pitching somewhat ahead of their audience's horizon of expectations.

Reality TV has nevertheless continued to push the boundaries. September 2007 for example saw CBS's launch of *Kid Nation*, a series in which 40 children (some as young as eight years old) were left to take care of themselves for 40 days in a ghost town in the New Mexico desert. (One has flashes of *Lord of the Flies* – if a version had been directed by Sergio Leone.) Others have gone even further: in September 2007 it was reported that the presenter of a Brazilian reality crime show had been accused of ordering contract killings in order to boost his programme's ratings.

Contemporary television at times is becoming, it appears, almost an inadvertent parody of itself. Another example of this shift into the realm of the absurd might be *The Naked Office*, launched by the UK's Virgin television in July 2009, a show in which a marketing company in Newcastle attempted to boost morale and productivity by introducing Naked Fridays – or, for that matter, *Autistic Superstars*, screened by BBC Three in May 2010, a well-meaning talent contest for young people with autism – or indeed in December 2011 when Valerio Zeno and Dennis Storm, the presenters of Dutch television's *Guinea Pigs*, ate surgically removed and sautéed sections of each other's flesh. Populist television seems to be consuming and regurgitating itself with the frenzy of a cannibalistic bacchanal.

When, however, reality television intentionally caricatures itself, the willingness of the public to accept the parody at face value further demonstrates the extremes and absurdities which this mediation of reality has reached – as well as its influence over public perception. We have increasingly come to accept the most absurd caricatures of reality television as real, and a number of hoax reality television shows illustrate this phenomenon. In November 2005, Britain's Channel 4 announced a programme called *Space Cadets* in which nine people believed they were in training in Russia to become cosmonauts (rather than, as it turned out, still in the UK on a mock training programme) and that the winner would be launched into space; in May 2007, in order to highlight a shortage of organ donors, Dutch television caused an international controversy when it screened *The Big Donor Show* – in which an apparently dying woman (in reality played by an actress) would choose from among three contestants (real patients in need of transplants) to receive her kidneys; in March 2010 French television faked a reality television show in which 80 people were asked to zap a contestant with increasingly powerful electric shocks each time he failed to answer a question correctly – and in which, urged on by the cheers and cries of 'punishment' from the studio audience, 64 of these 80 members of the general public, despite the contestant's screams and pleas for mercy, pushed the punishment to what they believed to be a potentially lethal voltage.

Within the depthless environment of early twenty-first-century mass media culture, these hoaxes (however eccentric, however far they pushed the limits of rationality and taste) were not only believable – they were actually believed – not only by their participants (as in the case of the British and French programmes) but also (in the case of the Dutch programme) by the most seasoned and sceptical media commentators who – prior to its broadcast – lined up to call for the cancellation of the show, their protests framed within a credulous rhetoric which repeatedly invoked the lamentable state of contemporary popular culture. Indeed, this author himself agreed to appear on air to denounce *The Big Donor Show*.

Terms of Estrangement

It was reported in July 2011 that for the first time the computer technology and media corporation Apple had more cash reserves than the United States government. The latest figures from the U.S. Treasury Department showed an operating cash balance of $73.7 billion, while Apple's most recent financial results boasted reserves of $76.4 billion. The following month it was announced that Apple had for the first time become America's most valuable company, with its market capitalization for the first time surpassing that of oil giant Exxon. The iPad has become bigger than the oil rig, the iPhone more robust than the United States Treasury. That a media organization could be richer than the world's richest nation, or for that matter than that traditionally oil-reliant nation's biggest oil producer, seems absurd; but it sometimes seems that the twenty-first century's post-industrial civilization can appear so consistently absurd that we barely any longer even notice it.

The month before the British general election of May 2010, for example, *The Guardian* newspaper reported that Labour Party strategists had decided to capitalize upon reports of the Prime Minister's volatile temper with a (literally) hard-hitting public relations campaign: 'Following months of allegations about Brown's explosive outbursts and bullying, Downing Street will seize the initiative this week with a national billboard campaign portraying him as "a sort of Dirty Harry figure". One poster shows a glowering Brown alongside the caption "Step outside, posh boy," while another asks "Do you want some of this?"' It was perhaps only the fact that this story was published on 1 April which made it clear that it was a hoax. In the world of contemporary mass media culture it appears difficult to be so unprecedentedly absurd as to be universally incredible.

New media appear at time to make people and organizations lose all sense of themselves, and of proportion. It is as if their distance and relative anonymity or unaccountability release the users from their senses of reality. We might select any number of media-related news stories to illustrate this phenomenon ... The report, for example, that the British Foreign Secretary had in June 2009 posted a tribute to the late Michael Jackson on Twitter:

'Never has one soared so high and yet dived so low.' The accounts in May 2010 of 'riots' erupting during delicate negotiations between British Airways and the trade union Unite after the union's General Secretary had leaked details of the meeting onto Twitter. The tales that (in January 2010) council members in Cornwall were reprimanded for sending Twitter messages during a council meeting – messages which mocked other councillors – and that (in March 2011) a Blackpool councillor had described his constituents as 'donkey-botherers' on Facebook. The story of February 2010 that the Conservative Party chairman had complained when a Labour MP described the Conservatives as 'scum-sucking pigs' on Twitter. The case of a Labour Party candidate banned from standing at the May 2010 election as a result of his announcing his penchant for drunkenness, describing the elderly as 'coffin dodgers' and posting a series of obscene rants about fellow politicians on Twitter – or of another Labour Party candidate (for the Welsh Assembly election of spring 2011) who posted onto Facebook that he hoped that former Conservative Prime Minister Margaret Thatcher would soon die – or (reported on the very same day in April 2011) of the Kent Conservative candidate who branded local women 'sluts' on Facebook – or even of the Liberal Democrat councillor who, in February 2009, had faced calls for his resignation after posting onto Facebook pictures of himself dressed in Nazi uniform at a fancy dress party. The news of February 2010 that the toy manufacturer Mattel was planning to release 'Puppy Tweets', a device to allow dogs to post on Twitter. Or even the report (of January 2010) of a man concerned that heavy snowfall at his local airport would spoil his travel plans who was arrested under the Terrorism Act, suspended from work and banned for life from the said airport after he posted on Twitter: 'You've got a week and a bit to get your shit together, otherwise I'm blowing the airport sky high!' Or the similar case (of January 2012) of the British man who was refused entry into the United States after having tweeted that he was going to 'destroy America' (the BBC report observed that 'he insisted he was referring to simply having a good time.') Or, surely most extraordinarily, the revelation in May 2012 of the sorry saga of the British Prime Minister who signed off text messages to one of the country's leading media executives with the abbreviation 'LOL' believing that it meant 'lots of love' rather than (as the nation did when it heard the tale) 'laugh out loud'. We may note that only the first of these stories turned out to be a hoax – yet it seems no less plausible than the rest.

In September 2002, when a woman called Rena Salmon murdered her husband's lover and then texted him to tell him she had done so, she added in that message that her statement was 'not a joke'. Williams (2003) has witnessed in this assertion the failure of contemporary society to be alert to authenticity – which is perhaps to suggest that (when, for example, acts of war come to look like Hollywood movies or video games) we have become so far removed in our daily lives from traditional verisimilitude that even in the moments of the most urgently immediate reality, we doubt the veracity of material existence: 'that surely cannot be a plane hitting that skyscraper,' news viewers might have mused, 'that must be a computer-generated effect.' By translating it into the medium of the text message or the tweet, the event becomes at once real and unreal: it enters into the new paradigm of a hypermediated reality, which is at once detached from material reality (and therefore from traditional moral responsibility) and is also the only reality there now is.

Through the exponential developments in their spread and speed of operation, new media at once authenticate and exacerbate the contemporary state of absurdity, transforming wild fantasy into perceived reality. Stories of entirely imaginary provenance have repeatedly appeared in the output of nationally recognized news organizations. For example, it was reported in September 2009 that two Bangladeshi newspapers had published an article taken from a satirical American website which claimed that Neil Armstrong had told a news conference that the Moon landings had been faked. In August 2011 it was revealed that reports from the BBC, CNN, *The Daily Telegraph* and the *Daily Mail* that users of Internet Explorer had lower IQs than people who used other browsers had been the result of an online hoax. The organization which had published the information (based on what it had claimed to be its own original research) had only been established – that is, it had only established a website – the previous month. The media organizations who picked up on the bogus organization's press release had not thought to question its authenticity. It was, after all, available online. The hoax organization posted onto its own website the announcement that it had been set up in July 2011 in order to launch a fake 'study' called 'Intelligent Quotient and Browser Usage.' It added that 'the study took the IT world by storm.' What this hoax actually demonstrated was the credulity of traditional media organizations in the face of new media's pseudo-authenticity.

Another surreal fiction – which originated as a viral email and, somehow legitimized by its online ubiquity, appeared on the websites of various news organizations in January 2005 – concerned the antics of a Slovakian man who reportedly urinated his way to freedom after his beer-laden car was caught in an avalanche. These stories are immaculate simulacra, copies without originals, relentless reflections which lack material objects. Just as media globalization dissolves geographical and cultural boundaries, so a hypermediated society's emphasis upon the virtualized contemporary moment rips narratives from their material history and cannot therefore any longer distinguish between those which have lost that history and those which never had it in the first place. As Facebook and MySpace have become sites for the generation and dissemination of that which is perceived as information, we have witnessed what Redden and Witschsge (2010: 185) have described as a blurring of the spaces of news and of non-news: a blurring of the virtual and material, of fantasy and history.

At the end of October 2008 the case of a Welsh road sign was widely reported in the British media. The Welsh language translation of the sign was supposed to say: 'No entry for heavy goods vehicles. Residential site only.' However, Swansea Council's translator was out of the office when the email requesting the translation arrived; as a result, the sign now says (in Welsh): 'I am not in the office at the moment. Send any work to be translated.' The medium is the message: that is, the mediated message no longer has any meaning left to it save for a delineation of the processes of the medium itself. The image (the translation) no longer refers back to an original object: this is the absurdity of the hypermediated simulacrum.

Just as these modes of mass-mediation absurdly frame our lives, so they tragically frame our deaths. In 2003 Sharon Osbourne's successful battle with cancer was documented by the second season of *The Osbournes*; in 2008 Jade Goody's less successful fight against the disease was portrayed in the final episode of *Living with Jade Goody* on the UK's Living TV. One need look no further for examples of this phenomenon than Sky One's

How Not to Die (2005), a former SAS officer's guide to surviving perilous everyday situations (complete with relentless simulations of various versions of the presenter's own death), or Spike TV's *1000 Ways to Die* (launched in 2008), a series which reconstructed entertainingly surreal 'real-life' deaths. Yet the processes of hypermediation far exceed such playful simulacra. They not only virtualize our deaths; in some cases, they may also realize them. In January 2007, for example, a 28-year-old woman called Jennifer Strange died as a result of drinking too much water whilst taking part in a Californian radio station's 'Hold Your Wee for a Wii' competition (whose goal was to drink lots of water, defer urination and win a digital game). In October of the same year Anthony Henderson was jailed for three years after having been filmed on a mobile phone urinating on a woman as she lay dying in the street and shouting 'This is YouTube material!' In March 2010 it was reported that a South Korean couple had become so obsessed with nurturing a virtual daughter in the popular role-playing game Prius Online that they allowed their own three-month-old baby to starve to death. The following month Minnesotan William Melchert-Dinkel was accused in court of encouraging the suicides of a British man and a Canadian woman by posing as a female nurse in online suicide chatrooms and advising people on how to take their own lives.

The convergence of new media forms and technologies (reality television, social networking websites, the home video camera) foster unprecedented conditions of cultural and social interaction whose rules of negotiation are as yet poorly understood and which may as a result catalyze unprecedentedly horrific and terrifying possibilities. In October 2009 Petros Williams made a 'farewell video' of his four-year-old daughter and two-year-old son as they watched *The X Factor* on television before strangling them with Internet cables. When his case came to court the following March prosecutors described his crimes as representing a 'symbolic act of punishment' against the children's mother for her use of Internet dating sites to converse with other men. *The Sun* newspaper reported on 23 March 2010 that the mother had moved out of the family home shortly before the killings as a result of 'rows over her use of the web.'

The fact that we cannot make any sense of this does not make it any the less real. Indeed the reason why we cannot make any sense of this at all is precisely the fact that it is real – but it does not fit within our traditional paradigms, the outdated paradigms of a pre-globalized, pre-virtualized (still heterogeneous, still material) world – nor within the expectations of a hyper-controlled utopian virtuality. The absurdity is not so much merely our current situation – the absurdity is that (because we still recognize traditional ideals of society, democracy and material history) we continue to try to apply our old logic to our new environment. The current state of global military conflict is a definitive symptom of this paradoxical condition: the absurdity of this war (a war with no end in sight insofar as it is a war whose *telos* is its process; a war fought to sustain obsolete notions of freedom and representative democracy; a war defined as being waged against such abstract and imaginary concepts as weapons of mass destruction, enemy combatants and terror itself) is epitomized by the startlingly irrational attempt to reconcile its parallel strategies of 'hearts and minds' and 'shock and awe' – new media communications strategies and *Blitzkrieg*. Stuart Price (2010: 188) has pointed out that the absurdity of the War on Terror leads us to conclude that it would be 'a mistake to expect it to follow a "rational" path.' Price suggests that this absurdity and irrationalism result from the contradictory motives underpinning the war. A congruent set of contradictory perspectives and paradigms underlies the new media age as a whole, and it therefore seems perhaps more appropriate to acknowledge (which is not the same as to accept) this absurdity rather than to attempt to rationalize it or to explain it away.

This study examines the strangeness of the world into which we appear to be heading: the social, political, economic and cultural paradigms fostered by the development of new media technologies, and the hegemonic responses to the resistance to these paradigms (at its most extreme, the War on Terror witnessed as a militaristic reaction to terroristic resistance to media globalization). In doing so, it explores the apparent paradoxes and absurdities which have arisen from the tensions between traditional and evolving world views. It investigates the influences of the uses of new media technologies in government, games, television, social networking and knowledge distribution upon conditions of personal, political and social subjectivity. It attempts to draw parallels between such apparently diverse media phenomena as online elections, militaristic video games, *The X Factor*, Facebook and Wikipedia – that is, it recognizes in all of these phenomena reflections of the contradictions provoked by crucial shifts in the defining paradigms of western civilization – shifts prompted by the rapid evolution of mass media, information and communications technologies.

The Digital Village

'Everything,' wrote Baudrillard (1988: 32), 'is destined to reappear as simulation [...] terrorism as fashion and the media, events as television. Things only seem to exist by virtue of this strange destiny. You wonder whether the world itself isn't just here to serve as advertising copy in some other world.' Yet surely we have *always* been mediated, *always* lived in a mediated world? History, society and even human consciousness itself all require networks of communication – languages, media – in order to exist. So what makes the situation so different today?

What makes all the difference, a world of difference, is the fact that today's globalized multimedia are coming to represent a single monolithic *omnimedium*. Stuart Hall (1980) wrote famously of the possibilities of negotiated and oppositional readings of media texts; but these active readings are, of course, only possible if we are exposed to a range of discourses, a variety of meanings. So what happens when we have become reliant upon – and indeed inscribed within – a single medium, an all-encompassing code? Like Jim Carrey in *The Truman Show* (1998), how can we see things clearly when we are part of the picture, when the medium is our entire world?

When examining the impact upon society of new paradigms, structures and technologies, there is perhaps no place better to look than speculative fiction. Popular science fiction has, for more than a century, explored the unfolding future of western civilization, and its contemporary incarnations continue to investigate these problems and possibilities. Now more than ever, these fantasies articulate the absurdity, virtuality and extremity of the

new reality. In, for example, the opening episode of the American science
fiction TV series *Caprica* (2009) – a prequel to the reimagined *Battlestar
Galactica* (2004–2009) – a young woman on a world wrought by ethnic
tensions is killed in a terrorist attack perpetrated by a monotheistic, fundamentalist and extremist religious cult. Her father's attempts to memorialize
her – to concretize and externalize her memory – at first through her reconstruction in a virtual world, and then by transferring that reconstruction
into the physical world, set off a chain of events which leads towards the
near extinction of human civilization. The similarly self-defeating nature
of western civilization's attempts to perpetuate its traditions and paradigms
seems central to the absurdity which the new media age is experiencing.

Within this context one particularly striking (although not particularly successful) television series was the 2009 reimagining of the 1960s
Cold War fantasy *The Prisoner*. In the remake a New Yorker – an electronic
surveillance expert – discovers himself in a surreal new world, a village at
the heart of a desert, a community of false bonhomie as superficial and
inescapable as Facebook, a global village closer to Baudrillard's desert of
the real than to McLuhan's utopian vision of informational mobility and
interactivity – Abu Ghraib meets Disneyland via *Lost*. This post-historical
protagonist cannot access the past; his memories are fragmentary and
incoherent. The last thing he can remember is an explosion in New York.
After the devastation of 9/11 only the absurdity of this hypermediated
reality and its simulacrum of freedom remain: 'there is no New York,' says
the leader of this new world: 'There is only the village.'

This village suffers a permanent condition of heightened paranoia in
which everyone lives in fear and in which such fear is publicly perceived
as 'guilt in disguise'. Everybody distrusts everybody else, and everyone
spies on each other, even parents on their children – and vice versa. This
surveillance society founded upon a self-sustaining state of terror offers
overt contemporary parallels: the West has reinvented the strategies of its
abstract enemy in order to apply them not only to that imaginary other
but also to its own populations. In *The Prisoner*'s village, names have been
replaced with numbers: humanity literally has been digitalized. The series
opens with the death of an old man known only as 93, recalling perhaps the
crash of United Airlines flight 93 on September 11 2001. Since that date,

contemporary history has become as digitalized as the village, converted into a sequence of numbers: 9/11, 7/7. This is the monolithic reality of the globalized village in a time of virtual terror.

The series's opening episode ends with a vision of the twin towers in the heart of the desert, a mirage of a lost world, a faded paradigm which can no longer be reached. In the programme's fourth episode cracks begin to form in the reality of the village – actual cracks, holes in the ground on which the village is built. Beneath the surface there is absolutely nothing, a void. This truly is Baudrillard's depthless desert of the real. The village is, as Jean Baudrillard (citing Borges) might have had it, a full-scale map of the real designed to map directly over the real until the real recedes and the map becomes real, the only real at least which remains, and eventually (as history diminishes) the only reality that there has ever been (Baudrillard 1994: 1). It is a virtual realm, which, like the electronic state, the digital game, reality television and the social networking site, reinforces and exacerbates the very problems and paradoxes it had purported to resolve. The village is, as such, a map of the multimedia mapping of the world, a model of the Internet itself, or of the world for which the Internet has become a metaphor and which at the same time is merely a metaphor for that network.

The primacy of this virtual map and its potential for disparity with the physical world became jarringly clear when, for example, in June 2010, a Californian woman called Lauren Rosenberg launched litigation against Google after having been hit by a car on a four-lane highway while using a pedestrian-friendly route recommended by Google Maps. According to her lawyer, Google had 'created a trap with walking instructions that people rely on. She relied on it and thought she should cross the street.' By doing so Ms Rosenberg fell victim to the distance and tension which still persists between the material and its map – or rather, insofar as the map had become her dominant paradigm, between the map and its material. The problem, some might argue, is not the map: it is the fact that the rest of the physical world has not as yet (but that Ms Rosenberg apparently had) fully recognized its authority. *The Prisoner* envisions a realm in which that authority goes almost entirely unchallenged, an imaginary landscape whose absurdity only one individual – the series' protagonist – appears to recognize.

The Prisoner's unreal village is a shallow dream, the simulacrum of, as its custodian explains, a 'new world' invented to replace the pain and complexity of material history – because 'the world is not a pretty thing when you look at it too close.' *The Prisoner* imagines an escape from contemporary reality, and, like the *Matrix* films, it addresses at once the unsustainability of the illusion and, conversely, the threat that the illusion may become overwhelmingly real, may become the only reality left to us. The final horror of *The Prisoner*, its final revelation, is that in its closing episode the series's protagonist, the village's lone voice of resistance, the eponymous prisoner himself, becomes the gaoler, the born-again zealot and custodian of its pseudo-utopian illusion.

Fantasy has traditionally offered itself as an allegory of, or satire upon, urgent contemporary concerns; but it may be that, at the extremes of history (when history reaches its own extremity at its borders with the fantastical, the post-material and the virtual) the fantasy space itself becomes almost indistinguishable from the historical. As the extraordinary renditions of fantasy approach those of fact, so fact becomes ever more fantastical.

CHAPTER 2

Electronic Government

The French sociologist Pierre Bourdieu (1991: 172) has argued that a lack of access among the general populace to the tools necessary for political participation has resulted in the concentration of political power as the province of a small elite. Although much has been claimed for the potential of new media technologies to promote democratic political participation, it remains unclear whether the application of these technologies in practices as apparently diverse as those of electronic governance, interactive entertainment and virtual socialization indeed offer the popular dissemination of the technological and cultural capital which Bourdieu sees as essential to the processes of democratization – or whether they in essence divert their subjects from such processes.

Bourdieu (2005: 62) has proposed that 'to make a decisive contribution to the construction of a genuine democracy [...] one needs to work towards creating the social conditions for the establishment of a mode of fabrication of the "general will" [...] that is genuinely collective [...] based upon the regulated exchanges of a *dialectical confrontation* [...] capable of transforming the contents communicated as well as those who communicate.' It appears that the homogenizing seamlessness of contemporary media technologies refutes the possibility of any such dialectical confrontation; and therefore, if we are to attempt to forge in the smithy of society the political consciousness of the populace, then it seems vital to ensure that this process is not illusory, is not a forgery or *fabrication* in a sense unintended by Bourdieu. Those who might see the potential of emergent information and communications technologies to foster a global village in which electronic technologies unify society's fragments (McLuhan 2001: 385) – or for that matter a return to the 'vibrant democratic intellectual culture of the eighteenth-century London coffeehouse'

(Keen 2008: 79; see also Doyle and Fraser 2010: 229) – might note that the celebrated Habermasian notion of the public sphere – 'the sphere of private people come together as a public' (Habermas 1989: 27) – seems, for Jürgen Habermas at least, no longer an option: 'the communicative network of a public made up of rationally debating private citizens has collapsed; the public opinion once emergent from it has partly decomposed into the informal opinions of private citizens without a public and partly become concentrated into the formal opinions of publicistically effective institutions' (Habermas 1989: 247).

These new technologies do not in themselves afford a return to a determining sphere of public debate for which we may have grown nostalgic (and which indeed we may have fantasized, or at least idealized – or, more pertinently, maintained as an ideal towards which we might aspire); on the contrary, they appear to offer an illusion of individual public agency (an agency in fact held by institutional bodies) which consigns the individual to an extended and ubiquitous 'private sphere' – as Zizi Papacharissi (2010) has famously called it. Indeed, Papacharissi (2010: 167) notes that while the concept of the private sphere may offer to describe the processes of civic interaction in post-industrialized societies one should not thereby mistake it, in itself, for 'a recipe for democracy'.

This view does not appear entirely to have reached the political classes. In June 2009 Britain's then Prime Minister Gordon Brown published an article in *The Times* newspaper in which he argued that Internet access for the entire population of the UK was an essential factor in a bid to secure the healthy economic and democratic future of the nation. Brown wrote:

> Whether it is to work online, study, learn new skills, pay bills or simply stay in touch with friends and family, a fast Internet connection is now seen by most of the public as an essential service, as indispensable as electricity, gas and water [...] Digital Britain cannot be a two-tier Britain – with those who can take full advantage of being online and those who can't.

In July 2010 Brown's successor David Cameron backed the Networked Nation Manifesto, a report produced by the government's Digital Champion Martha Lane Fox which announced the need to get all British people of working age online by 2012 (a goal which does not appear to have been reached): 'digital inclusion [Cameron stressed] is essential for a modern dynamic economy.' Brown and Cameron did not go quite so far as France's Constitutional Council which had, in June 2009, ruled that Internet access – as 'an essential tool for the liberty of communication and expression' – represented a fundamental human right, as laid down within the Declaration of Human Rights in the preamble to the French constitution. Both, however, shared the notion that access to the Internet must be available to all in order to maintain conditions for the ongoing development of a sustainable and equitable modern state, if (as seems inevitable) that technology is to define the industrial, social and political structures of the globalized future.

But has such equity as yet been achieved in those parts of the world which have led the way in the development of the digital state (let alone on the rest of the planet)? Does digital inclusion promote democracy – or might digital exclusion and imbalance undermine the prospects for democracy in a contemporary society?

In their study of the uses by members of the U.S. Congress of online social networking technologies for establishing dialogues with their constituents, Glassman et al. (2010: 11) note, for example, that such online dialogues might become influential in the determination of legislative voting and policy decisions. In his study of similar systems in the British Parliament, Williamson (2009: iii) however notes that while new technologies can improve communication flows with constituents it remains the case that digital media do not represent an unproblematic panacea. Clearly one of the ongoing problems with these tools is that they remain in the hands of the computer-literate and Internet-active section of any electorate; and that it is conversely those constituents who lack this intellectual and technological capital whose situations may most urgently need to be considered in the development of policy agendas.

This is not, for example, to disparage the hopes of such theorists as Coleman and Blumler (2009): aspirations that new media technologies may eventually play a part in restoring a disillusioned public's trust in the structures and institutions of democratic politics; but it is to emphasize that these technologies cannot achieve such a goal in themselves – and that, if we assume they can, opposite tendencies may emerge.

The New Democracy

A former senior engineer at the website, Karel Baloun (2006: 88) has asked whether Facebook will 'somehow change the world, starting with its political election-focused groups and candidate profiles.' What is perhaps most interesting about this question is the fact that it is being asked: that fact that we might seriously be expected to take the political impact of such a social networking site quite so seriously.

In April 2010, four weeks before a British parliamentary election, it was reported that the UK Electoral Commission had collaborated with Facebook to encourage unregistered voters to enfranchise themselves. Facebook users visiting the site were asked if they had registered to vote, and if they had not, were directed to a page that would allow them to register online.

Can the use of such social networking sites as Facebook, then, promote participation in democratic processes? And should it do so: should we be specifically targeting those with Internet access in voter registration? On the face of things, such high profile political endeavours as Barack Obama's electioneering pages on Facebook would seem to suggest that such sites can foster democratic activity. There remain, however, significant questions as to whether the fundamental ideological perspectives underlying and advanced by these homogeneous media forms are themselves consistent with western notions of democracy, diversity and socio-political engagement.

On 15 April 2010, immediately after the UK's first ever televised election debate featuring the leaders of the country's three main political parties, Rory Cellan-Jones, *BBC News*'s Technology Correspondent, reported on his Twitter page the news that 36,483 people had posted a total of 184,396 messages onto Twitter in relation to (and during) that debate. In an article on the *BBC News Interactive* website the previous day Cellan-Jones had pointed out that Facebook would meanwhile run its own digital version of the leaders' debates – and that all three party leaders had 'agreed to answer questions submitted by users.' The morning after the first debate, on the BBC's *Breakfast* news programme, Cellan-Jones added that Facebook had experienced a server overload as a result of online activity related to the debate. Yet Cellan-Jones also suggested that this activity did not represent a step change in political participation – it had merely shifted extant offline debates onto these sites. One might, however, ask whether the limitations and structures of these sites allow for the range, extent and freedom of debate possible offline – or whether they offer a regulated imitation of and replacement for such participation practised in the non-virtual public sphere. Answering questions through Facebook (or indeed allowing one's advisers to do so) seems rather less onerous a task for a politician than facing the public and their political opponents in a live television debate – or facing the opposition in parliament – or facing journalists in interviews or press conferences.

Peter Bradshaw of *The Guardian* newspaper commented to the BBC in April 2010 that the UK election campaign (then ongoing) had witnessed the sudden collapse of the traditional press power to set the news agenda: 'these reality-TV-style debates have overnight completely changed the political landscape.' The point is not simply that these debates moved political agenda-setting into the realm of television (television had shared this role with the press for some decades), but specifically that this process had been appropriated by an interactive multimedia format which owed much to the style and structure of reality television. Furthermore, these televised election debates had at once blurred the distinctions between, and in doing so drawn attention to the tensions between, older and newer media paradigms: between, for example, the conflicting etiquettes of broadcast television and narrowcast online discourse. In the immediate wake of the first of these live leaders' debates, it was, for example, reported that ITV news had accidentally broadcast an obscene message abusing Conservative Party leader David Cameron: a shot of the readers' discussion page on the station's debate webpage which focused on a particular post which read: 'Cameron – You're a first class cunt.'

The potential entertainment value of the interactive multimedia circus which had grown out of this democratic process did not go unnoticed by commercial interests. During Britain's general election campaign of

April–May 2010, the company producing the yeast-based spread Marmite created parallel Facebook sites for the Marmite Love Party and the Marmite Hate Party, complete with slogans ('Spread the Love' and 'Stop the Spread' respectively) and campaign statements, manifestos, pledges and profiles of the party leaders – as well as links to the Marmite News Network which kept readers updated on the latest campaign news. Users with strong positive or negative opinions on the product could log onto Facebook to register their vote for or against the spread. By the end of the election period more than 150,000 people had registered their support of the Marmite Hate Party's campaign – while the Marmite Love Party had attracted over 350,000 supporters. One is reminded of the 2009 campaign launched in Facebook to secure a number one Christmas single for Rage against the Machine as a protest against *The X Factor*'s dominance of the yuletide music charts, a campaign which prompted more than half a million Facebook users to register their support. Is this then the extent of political engagement offered by Facebook – an illusion of engagement which, like that offered by the video game, at once sublimates and trivializes the desire for socio-political agency?

In April 2010 it was reported that the manufacturers of Marmite were threatening to take legal action against the extreme right wing British National Party in order to ban its use of an image of a jar of the yeasty spread in a political broadcast. Marmite's spoof campaign had itself been hijacked by the anti-democratic extreme of British political opinion; the parody of democratic processes had become a tool for forces actively seeking to undermine those processes. This seems an appropriate, if absurd – that is, an appropriately absurd – model for the effects of hypermediation upon the processes of politics.

Everyone who is anyone in public life has to have a Facebook page. The British royal family launched theirs in November 2010, shortly after George W. Bush had launched his own, but many public figures (from Barack Obama to Gordon Brown) have tried (with varying degrees of success) to exploit the site for rather longer. By November 2010, six months after his election to office, British Prime Minister David Cameron, for example, had over 92,000 fans of his Facebook page. It may be noted however that at the same time the Facebook page entitled 'throwing eggs as David Cameron, brick shaped eggs – made from brick' had over 99,000 fans, while 'David Cameron is the new PM. Last one out of the country turn off the lights' boasted over 108,000 fans and more than 204,000 people supported the page entitled 'David Cameron wants change! Give him 30p and tell him to f**k off.'

One might argue that, if political participation through Facebook or other online activities is meaningful and empowering, then such activity biases democratic processes in favour of those groups in society (the economically, technologically and educationally advantaged) who least need that empowerment. Conversely one might assert that, if such participation merely offers an illusion of empowerment and socio-political agency which trivializes and denigrates politics itself, then it in fact undermines the desire, and therefore the potential for, real empowerment and meaningful agency. The Internet user feels as though she is at the centre of the universe, and this imaginary repositioning pre-empts the subject's struggle towards such centricity and significance.

In the immediate wake of the British general election of 6 May 2010 – which resulted in a hung parliament and emergency talks between the Conservative Party and the Liberal Democrats over the possibility of a coalition which would make the formation of a government possible – a Facebook group entitled 'We don't want the Liberal Democrats to make a deal with the Conservatives' was established. According to the founders of the group, 'the Lib Dem head office advised us to start a group to get an idea of how many people object to the proposed coalition.' After five days this group had over 60,000 members – a figure dwarfed by the support shown to Marmite and Rage against the Machine, but still perhaps a not entirely insignificant representation of the electorate (nearly one per cent of those who voted for the Liberal Democrats at the election). A similar Facebook group named 'We don't want the Liberal Democrats to make a deal with Labour' was also started: within the same time period it had attracted a total of 83 members. In addition to this a Facebook page called 'We don't want the Liberal Democrats to make a deal with the Labour Party' had attracted 203 supporters and another page called

'We don't want the Liberal Democrats to make a deal with Labour' had gained 59 supporters. It therefore appeared that those against a pact with the Conservatives very heavily outnumbered those opposing an agreement with Labour. At the end of the fifth day after the election, however, the Liberal Democrats abandoned any possible agreement with the Labour Party when they announced a deal with the Conservatives. So much then for the power of Facebook: so much perhaps for the Internet's capacity to channel public opinion directly to the political elite – so much, at least, for that elite's willingness to take that opinion on board.

It may, however, not only be the opportunities for political participation which such new technologies are subverting. As we shall see, it may be that these new technologies, in their radical reshaping of the conditions of social subjectivity and of social relationships, are eroding the very possibility of the traditional western model of politics itself.

Digital Delivery

The availability of the electronic provision of commercial services (in, for example, the financial industry or the information supply sector) is virtually ubiquitous in the post-industrialized world, and the replication of this model in the public sector (in the form, for example, of online tax returns, direct debit digital payments for state services and the electronic dissemination of public service information) is similarly common. As Coleman (2005c: 207) suggests, as commercial activities have been radically reshaped by these new technologies, it would have been unlikely that processes of political representation would not also be influenced by these trends. Thus the commercial e-service model has been translated into the area of e-government. The successful adoption of this paradigm by political parties (such as by Barack Obama on his journey to the White House) is one manifestation of this phenomenon. Barack Obama's first presidential campaign has been described by Papacharissi (2010: 1) as having achieved a degree of 'media-savvy notoriety' for its innovative deployment of new media technologies. The Obama model involved not only online fundraising and campaigning activities, but also the possibility (or at least the semblance) of participation in policy debates (on, for instance, the candidate's Facebook page) – although it is uncertain to what extent (if at all) these online debates might have framed the presidential agenda. A number of governments have made similar use of (or claims for) the Internet as a site for democratic participation. Proponents of electronic politics and electronic government have argued for their potential to move beyond the commercial sales-and-supply model towards a two-way or dialogical (or multi-path) paradigm. E-government's advocates announce the possibility of something more than e-services; they suppose that it (not unattractively) offers the chance of e-democracy – of interactive popular participation in political processes: 'interactivity opens up unprecedented opportunities for more inclusive engagement in the deliberation of policy issues' (Coleman 2005c: 209). This is a vision in which, as Coleman and Spiller (2003: 11–12) propose, 'interactive technologies, which can facilitate online consultation and dialogue, make direct representation a possibility.'

But this is only the beginning. If democracy's ultimate expression and test take place at the ballot box, then electronic democracy may be seen as coming of age in the introduction of online voting. Stephen Coleman (2005b: 97) writes that, while online voting may be take place in traditional polling stations, it can also be employed, more ambitiously and radically, as a means to afford voters the opportunity to participate in elections from home or at work or wherever they can access the Internet.

Risks of course accrue to this process. Coleman (2005b: 96) pertinently enquires as to the nature of these risks: 'Are risks purely technological or are there other processes at stake?' As he goes on to suggest, these risks (which technologists may see merely in terms of systems security) may become matters of *moral* security – when for example, voters without Internet access might consider the limitations upon opportunities for online voting as representing 'an act of discrimination' (Coleman 2005b: 99). This chapter addresses these issues of moral security: other studies (for example, the work of OSCE) more closely examine – and for the most part legitimize – the technological security aspects of online voting.

Some argue that those moral risks are outweighed by the possible benefits for whose sake those risks might be incurred. Those perceived or projected advantages tend to based upon ambitions of increased civil and democratic participation prompted by the uses of these technologies: that these technologies might offer 'a counterbalance against voter apathy and therefore increase voter turnout, which in turn legitimizes the outcome of the electoral process' (Xenakis and Macintosh 2007: 14).

However, as Xenakis and Macintosh go on to stress, this hypothesis remains unproven. In, for example, the local elections of August 2007 the British Electoral Commission ran a series of pilot schemes involving online and telephone voting services in the districts of Rushmore, Sheffield, Shrewsbury and Atcham, South Buckinghamshire and Swindon. The Electoral Commission's subsequent reports agreed that the schemes had 'a negligible impact on turnout.'

The Reinforcement Hypothesis

The ethical, educational, social and political ramifications of the sudden proliferation of new information and communication technologies have in recent years become a matter of increasing concern to scholars working in the fields of computing, sociology, media studies and politics. In their study of *Computer Ethics and Professional Responsibility*, Bynum and Rogerson (2004b: 318) address the crucial question as to whether the Internet will sponsor global democracy or become a tool for the manipulation of the general public by political elites. Bynum and Rogerson (2004a: 6) describe a contemporary scenario in which academic opinion on these questions is polarized between cyberoptimism and cyberpessimism:

Optimists point out that information technology, appropriately used, can enable better citizen participation in democratic processes, can make government more open and accountable, can provide easy citizen access to government information, reports, services, plans, and proposed legislation. Pessimists, on the other hand, worry that government officials who are regularly bombarded with emails from angry voters might easily be swayed by short-term swings in public mood [...] that dictatorial governments might find ways to use computer technology to control and intimidate the population more effectively than ever before.

In April 2012, for example, *The Independent* newspaper ran as it front page story the news that the British government were introducing legislation to afford police and security services the power 'to watch you on the web' – 'to check on citizens using Facebook, Twitter, online gaming forums and the video-chat service Skype.' The paper took some relish in retrospectively quoting David Cameron – a year before he had become Prime Minister – railing against the UK's surveillance society and arguing that 'we are in danger of living in a control state.' The following day, writing in *The Sun* newspaper, Home Secretary Theresa May attempted to reassure the British public that 'no one is going to be looking through ordinary people's emails of Facebook posts' – only those of serious criminal suspects. One is tempted to ask how Ms May would define 'ordinary people'. The same day *The Sun* newspaper featured comments from another senior Conservative politician warning that this development might turn Britain into 'a nation of suspects' – and from the government's Information Commissioner suggesting that this move represented a 'step change in the relationship between the citizen and the state.'

The risks of electronic governance may often appear clearer than the benefits. The political scientist John Curtice (2009) has pointed out that 'people's uses of the Internet are primarily a function of their prior motivations. The Internet isn't bringing into political activity people who aren't engaging in it offline.' Papacharissi (2010: 105) has argued that 'digitally enabled civic activity has not been associated with an increase in political participation [...] nor has it been identified as a factor in reducing voter cynicism and apathy.' Gibson et al. (2004: 8) suggest that it is highly questionable whether e-government has any effect on democratic participation. As

Nixon (2007: 29) points out, research studies have suggested that between 60 and 85 per cent of e-government projects can be regarded as failures. More problematic, however, than the mere failures of new technologies to foster civil society and democratic participation are suggestions that these systems may in fact undermine opportunities for equitable popular representation.

As well as increasing public access to information, new media technologies may, for example, allow governments and economic interests greater powers to disseminate and legitimize their own agendas. As Catherine Needham (2004: 65) notes, governments can uses processes of direct communication and consultation with their electorate to sideline the powers of elected legislatures to hold those governments to account. In their study of Internet use in Russia, for example, Fossato and Lloyd (2008: 56) argue that while liberals may see new media technology as a tool for individual liberation its potential, in Russia at least, may primarily become focused upon social manipulation. Such scenarios have led the growing ranks of cyberpessimists towards what has been dubbed the reinforcement hypothesis. This theory is neatly summarized by Raab and Bellamy (2004: 21) when they argue that 'technology becomes a tool for the reinforcement of existing power structures.'

Martin Hand (2008: 77) has noted that western governments have grown increasingly interested in the use of digital technologies as tools for those governments' own reinvention and legitimization. One might suggest that any such reinvention appears to have been focused upon the entrenchment of power hierarchies rather than the opening up of government to greater transparency, accountability and popular participation; and one might also observe that the emphasis of most governmental initiatives in the deployment of these technologies has been concentrated upon the legitimization of extant institutions, structures and practices of power.

The uneven distribution of new media technologies' tools of access to intellectual and cultural capital is a problem not only for an electronic society as a whole; it is also a specifically critical issue for the practices of online government. Indeed, in 2003 the United Nations's World Public Sector

Report warned that broad sections of national populations were not necessarily reaping the benefits of state investments in electronic governance.

Neumann (2004: 208) pertinently asks whether we are creating a bipolar society split between those who have access to information technologies and an increasingly disempowered and disenfranchised sector of those who do not. According to Margolis (2007: 2) the electronic gap has come to reflect the socio-economic, educational and demographic divides of the non-virtual world, as the political and economic structures of cyberspace echo those found in the material world. As Bolter and Grusin (2000: 182) argue, access to these technologies confers a 'conspicuous social status' of some significance in the post-industrial world.

Nixon and Koutrakou (2007: xxi) report that individuals with lower levels of educational achievement are less likely to use the Internet. This not only leads to a reinforcement of social, economic and educational schisms forged along demographic lines, but may also imply a set of specific and acute risks incurred by the adoption of new communication technologies for the purposes of political, parliamentary and governmental information, consultation and representation. As Nicholas Pleace (2007: 69) points out, within any nation sectors of society with low levels of Internet access tend to correlate with those sectors which suffer low levels of household income (those very sectors which most need the engagement and support of the state): 'The implications for electronic voting are obvious. The poorer parts of the population are less likely to vote.' It may therefore appear inevitable that, as Åström (2004: 107) suggests, online voting 'shifts the bias toward the middle and upper classes: the already politically active.'

In an interview with Chris Middleton (1999), the IT ethicist Simon Rogerson proposed that 'the concept of online government implies literacy, and an awareness and acceptance of technology. That isn't the case in practice.' In this context, it is perhaps helpful to examine the case of a country which has in recent years stood at the forefront of developments in electronic democracy, government, industry and commerce: the small, modernizing and relatively new north-eastern European nation of Estonia.

E-stonia

Perched on the north-eastern edge of Europe, and wedged in between
Finland, Russia and Latvia, the Baltic republic of Estonia achieved inde-
pendence from the Soviet Union in 1991, and joined the European Union
on 1 May 2004.

Stephen Coleman (2005a: 6) has commended the Estonian government for its practical uses of the Internet to foster democratic participation.
Pratchett (2007: 10) similarly cites Estonian examples of major e-democracy
initiatives. Much has been made of Estonia's leap into the electronic age,
perhaps because its rapid and massive adoption of new media technologies
may be seen as affording a bridge from its past as a minor republic of the
Soviet Union, annexed by Stalin in 1940, towards its current situation as
an independent democracy based on the principles of market economics,
and as a member of NATO and of the European Union. In many ways
Estonia has represented to various commentators an admirable model for
the development of liberal democratic nationhood in the wake of political
oppression and a command economy.

Since the mid-1990s new technologies have undoubtedly sponsored
the internationalization of Estonia, in terms of its economic, industrial,
cultural and educational development. Its middle classes have enthusiasti-
cally adopted the systems and perspectives of western capitalism, and the
majority of its population has benefited, materially, socially and politically, from these new structures. The prevalent assumption, however, that
new media technologies have at the same time directly underpinned the
country's processes of democratization (in terms, for example, of electronic governance and online voting) is rather more open to question.
Indeed one might suggest that the uses of such technologies may in fact
have undermined Estonia's democratic development, in that they may
simultaneously have reinforced commercial and political structures and
entrenched socio-ethnic divisions.

Estonia has a population of about 1.3 million people. According to
the Organisation for Security and Co-operation in Europe (OSCE 2007:

10) Estonia's population includes 125,799 people who are officially state-less. The vast majority of these non-citizens are ethnic Russians. Despite OSCE's recommendations that Estonia facilitate citizenship for this sector of the populace (OSCE 2007: 1), progress remains painfully (one suspects, grudgingly) slow: indeed, the Estonian government itself has admitted the failure of its own Population Ministry's ethnic minority integration policies (Tubalkain-Trell 2008). This disenfranchised minority represents the most extreme example of the massive disparity between Estonia's socially, economically, educationally and politically privileged and disadvantaged classes. Estonia has one of the European Union's largest gaps between the rich and the poor: 18 per cent of its population has a disposable income less than 60 per cent of the national median income (Mikecz 2005: 154).

Towards the end of 2007 Estonia was ranked by the Foreign Policy Globalisation Index as the world's tenth most globalized nation (Collier 2007). A crucial factor in this ranking was the country's assimilation of new information and communication technologies. Estonia is therefore a key case in the debate as to whether new media are likely to impove the lives and rights of entire populations, rather than merely of their more privileged sections – not only because of the country's current uneven distribution of economic opportunities and social and intellectual resources, but also because of the fact that Estonia has, for the last decade, attempted to position itself as a society at the cutting edge of contemporary information technologies: 'E-stonia' (as it has liked to fashion itself). It appears that the undoubted economic benefits achieved by Estonia's enthusiasm for electronic commerce and the IT industry may not be mirrored by the more problematic consequences of its ongoing experiments in the areas of electronic government and e-democracy.

In 1997 Estonia launched the 'Tiger Leap Initiative', an ambitious and remarkably successful programme of economic, technological and educational development designed to establish Estonia as a modern and competitive e-state. Even before this initiative, Estonia's financial industry had started to exploit new technologies. The world's first online banking services had, for example, started in 1995. By the end of 1996, there were only about 20 such services worldwide – and three of them were Estonian.

In 2003 a World Economic Forum Report ranked Estonia in eighth position (out of 82 countries) for its methods of putting the Internet to practical use. It came in at number three for e-government, and at number two for Internet banking. According to a UNHCR report (2009), the proportion of Estonians who choose to bank online has risen to 83 per cent.

Estonia's IT advances have inspired a certain amount of journalistic and academic hype. In 2004 *The Guardian*'s Ben Aris described Estonia as 'the most intensely wired country in the world.' At the end of April 2005, the Economist Intelligence Unit's sixth annual ranking of the 'e-readiness' of 65 different countries – in which Denmark, the United States and Sweden took the top three spots – placed Estonia at number 26, just ahead of fellow new EU member Slovenia.

According to the International Telecommunication Union, Estonia reported 33 per cent of its population as Internet users in 2003 – a figure two per cent higher than Ireland's (Fuller 2004). That same year, TNS Emor pollsters reported that as many as 45 per cent of Estonian residents aged 15 to 74, or around 576,000 people, used the Internet, and that this figure had risen by 59,000 since the previous year. More recently, a UNHCR report of 2009 suggested that Internet penetration had risen to 64 per cent.

In 2002 Estonia established a non-profit organization called the E-Government Academy. In an interview with *BBC News*, Ants Sild, a programme manager at the E-Government Academy, said: 'What we mostly teach [...] is understanding what your goals are as a government, and then figuring out how technology can help you achieve those goals' (Boyd 2004). One is forced to wonder in this connection whether an independent organization should in any way be defining the goals of government, and for that matter whether the goals of government are always necessarily consistent with the interests of its electorate.

In 2004 a ranking of the global development of e-government published by researchers at Harvard University placed Estonia in fifth place (Boyd 2004). The European Commission's fifth annual survey of online government services in Europe, released in March 2005, showed that, while most of the European Union's then ten newest member states scored in the lower half of the table – with average levels of e-government equivalent to those enjoyed by the EU's older member states in 2003 – Estonia alone of the new members appeared in the upper half (European Commission 2005). Indeed, according to Ernsdorff and Berbec (2007: 171), 'Estonia stands as the e-government leader in Central and Eastern Europe and as third in the world in e-government systems.'

The Estonian government has boasted the introduction of paperless cabinet meetings and claims that this measure has saved 100,0000 U.S. dollars a year in photocopying costs alone (in addition to a 200,000-dollar annual saving made by inter-ministerial memoranda going exclusively online). The government has also announced that their paperless condition has been responsible for reducing cabinet sessions (which had formerly lasted for up to 12 hours) down to a more manageable average of 45 minutes – or even, in one case, down to 14 minutes. As the government proudly proclaims these figures, one might however ask to what extent a quarter-hour cabinet meeting fosters the processes of mature democracy, administrative accountability and detailed debate which one might expect from the heart of a modern European government.

Since 1 January 2007, this craze for paperlessness has also come to encompass the publication of legislation. From that date the official government organ ceased its print publication (save for five paper copies) and now exists in an almost exclusively electronic form. For Marshall McLuhan (2001: 91) the adoption of the alphabet by European cultures allowed for a sudden and massive expansion of public literacy; the resultingly literate civilization afforded a structure in which all individuals had direct access to – and therefore were equal before – a written code of law. Estonia's decision to consign legislative and governmental records to a domain to which more than a third of the population lacks access might be seen as reversing McLuhan's politically progressive process. A system in which access to legislation requires degrees of technological expertise and facilities that are not available to the entire population appears no longer to meet McLuhan's basic conditions for modern civil society.

Another of the more obviously controversial aspects of Estonia's adventures in virtual administration was established in 2001 as the flagship project of its system of e-government. *Täna Otsustan Mina* ('Today I Decide') – or 'TOM' – was a website on which Estonian citizens could present proposals for legislation. If a proposal received sufficient support, it was to be discussed by the government. Although the portal boasted some 7,000 registered users, there were within a few years of its launch only 10 or 20 active members. The TOM portal prompted a number of minor changes in Estonian legislation and governance. One of these was a proposal to put the clocks forward in the spring and back in the autumn. Another was an amendment to the law on the possession of dangerous weapons – an exemption which permitted students of Tartu University (the country's oldest and most prestigious seat of learning) to carry swords on ceremonial occasions. It is notable that the first of these was enacted in order to bring Estonia into line with the time zone of its richer northern neighbour, Finland, a nation to whose prosperity Tallinn's middle classes self-consciously aspire. It may be seen not only as reflecting a concern of the country's economically advantaged internationalists, but also in the specific interests of the commercial enterprises in the wealthy area of north-western Estonia which do much business with the Finns. Furthermore, it virtually goes without saying that a focus upon the ceremonial sword-carrying rights of students at the country's elite university hardly provided evidence for this portal's stated intention to provide top-level governmental access for the Estonian people en masse.

The TOM portal received 359 proposals in the first six months of its operation, but by 2005 received only 49 proposals for the whole year (Ernsdorrf and Berbec 2007: 176). In fact, the website was generally considered an object of national ridicule, embarrassment or indifference. Mart Parve, technology correspondent for Estonia's most popular daily newspaper, *Postimees*, has called the initiative farcical, and has characterized its regular users as freaks and geeks: 'Some famous freaks are trying to start new laws, but it's not working. We've been quite pessimistic about the state since Soviet times. E-government is not employed at the level it should be. We're very interested in new technologies, but we don't use them properly.'

Estonia's TOM website eventually found itself sidelined and a new portal became tacitly acknowledged by the country's political elite as the primary arena for the proposal and discussion of legislation (Ernsdorff and Berbec 2007: 177). This website is run by an independent body, the Estonian Law Centre Foundation. Public opinion and legislative influence were now mediated neither by parliament nor even by a public or publicly accountable organization. Estonia appears to be leading the way in this new Europe: the laissez-faire diminution of the public sector is balanced by the increasing domination of the private.

On 4 June 2008, the Estonian government launched the *Osalusveeb* (Participation Web) as its successor to the TOM portal. According to the government website, this site would offer a chance for 'everyone [to] make suggestions to the state for simplifying public services'. With such an agenda set in advance, it seems unlikely that this site will offer significant opportunities for the Estonian people to formulate policy alternatives to the principles of public sector downsizing so essential to the permissive market economics of its centre right administrations. This new forum for debate appears to be telling the Estonian electorate that their civic duty (and the extent of their democratic privilege) is to devise policy details appropriate to the government's own political programme.

Electronic Democracy

In May 2005 the Estonian parliament passed legislation to introduce online voting at the country's local elections that October. According to *The Baltic Times* newspaper, 'Estonia would become the only country in the world where people could vote through the Internet from home. Although online voting is widespread in other countries, a voter must conduct his E-vote at a polling station computer.' The Associated Press added: 'Voters will need an electronic ID card, an ID-card reader and Internet access [...] It is estimated that nearly 1 million of Estonia's 1.4 million residents already have an official electronic ID card. The ID cards, launched in 2002, include small microchips and offer secure e-signing through a reader attached to their computers.' It should be noted that it is estimated that nearly a quarter of Estonia's population do not hold these ID cards, and it is believed that these people are mainly Russian-speaking residents.

According to Aleksei Gunter, a leading journalist at the Estonian newspaper *Postimees*, 'at the 2005 local elections, most of the e-votes went to the Reform Party. That was somewhat predictable, because the young and the well-off, who obviously have the means and the interest to use new technologies, favour that party.' In fact, according to the government's own report on that exercise in e-democracy, the Reform Party gave out ID card readers to their supporters during the election campaign (Madise et al. 2006: 41). The report of the Estonian National Electoral Committee (2007) shows that nearly 35 per cent of the electronic vote went to the Reform Party – significantly higher than the 28 per cent which it achieved in the total vote. As Ernsdorff and Berbec (2007: 178) report, 'some political parties considered e-voting an opportunity to increase their support, while others conceived it a threat.' Indeed, the then President of Estonia, Arnold Rüütel, attempted to veto the legislation which permitted online voting, on the grounds of its inherently inequitable nature and its openness to fraudulent manipulation, but was eventually overruled by the country's Supreme Court.

On 28 February 2007, Estonia extended e-voting to its national parliamentary elections. In March 2007 the BBC announced that 'Estonia has become the first country to use Internet voting in parliamentary elections.' *The Baltic Times* newspaper added that 'in what was hailed as the world's first full-scale Internet election, a total of 30,243 voters chose to log their votes online' (Alas 2007a). The total number of votes cast at the election was 550,213 and the total number of eligible voters circa 940,000. Ernsdorff and Berbec (2007: 171) have written that Estonia 'is setting an example in e-democracy throughout the European Union, being the first country in the world to enable all its citizens to vote over the Internet in political elections.' One might, however, call into question Ernsdorff and Berbec's use of the word 'all': although a majority (but certainly not all) of its citizens have Internet access, a much smaller proportion enjoy home access to the technology required to vote online.

It has been estimated that nearly two-thirds of Estonians currently use the Internet, although only about half of those have web access in their own homes. These people represent, for the most part, the country's educated urban middle class. The implementation of online voting has meant that this sector of the population (i.e. those who tend to be supporters of the centre right coalitions that invariably govern Estonia) will find it easier to exercise their democratic rights than those on the other side of the digital divide.

Voter mobilization is, of course, a key factor in the winning of elections (see Oberholzer-Gee and Waldfogel 2005: 74; Trost and Grossmann 2005: 128). In facilitating the voting process for their typical supporters, Estonia's centre right parties were therefore afforded an obvious electoral advantage by this mode of electronic democracy. This is a hypothesis with which even the former Communications Adviser to the Prime Minister, Tex Vertmann has 'theoretically' agreed. Ülle Madise (2007), Director of Audit at Estonia's State Audit Office, has claimed that online voting offers no advantage for e-voters – but, if that indeed were the case, one wonders why the state would bother with the trouble and expense of it at all. As Trechsel (2007: 37) points out, nearly 86 per cent of Estonians who chose to vote online did so because they found it more convenient than by traditional methods.

In an otherwise remarkably optimistic essay, Ernsdorff and Berbec (2007: 178) admit that 'e-voting has never been the result of popular demand but rather a result of the imposition of yet another initiative by a young Estonian political elite.' This is a position with which Jaak Aab, Estonia's Social Affairs Minister at the time of Estonia's first experiments in electronic democracy, would strongly disagree. Aab, who held this ministerial portfolio from April 2005 to April 2007, has commented:

> I believe that online voting encourages people to take an active part in democracy, because it gives an additional possibility to vote. It is especially important for people who cannot or have lower motivation to go to vote to the designated voting place. I am not concerned that it may primarily encourage participation among the educated middle classes, because Estonia doesn't have a big gap in Internet use, as lots of other European countries do. Research shows that in spring 2006, 60 per cent of Estonians (aged 6 to 74) were using the Internet. The Estonian government's aim is to provide the Internet to all Estonian people.

However, there appear to be several problems with Aab's argument. His first assertion – that online voting promotes democratic participation – seems to contradict the evidence of various empirical studies. Gibson et al. (2004: 3), for example, cite research which suggests that no modes of Internet use have been found to have any significant effect on individuals' tendencies to engage in politics.

Aab's dismissal of the education gap and the technological divide as irrelevant to Estonia is extraordinary, not only in the light of the country's massive socio-economic divisions, but also because it specifically contradicts the findings of his government's own report on the 2005 elections: that there were more people with high levels of education among e-voters (Madise et al. 2006: 30). Breuer and Trechsel's *Report for the Council of Europe* (2006: section 7) on the 2005 elections meanwhile found that Estonia's e-voting opportunities proved relatively unattractive for the elderly, for those with limited IT skills and for minority language speakers. We may note in this context that in 2010 the United Nations Committee on the Elimination of Racial Discrimination noted the low level of political participation of ethnic minorities in Estonia. (Indeed the OSCE report on Estonia's parliamentary elections of March 2011 observed that, in contravention of OSCE requirements, Estonia's 'long-term residents with undetermined citizenship do not have the right to join political parties.')

The *Report for the Council of Europe* on the 2007 election noted that 'the share of highly educated voters was almost 20 percentage points higher among e-voters than among traditional voters' (Trechsel 2007: 43). Trechsel's report stresses that 'the highest-income category is heavily overrepresented among e-voters' and that 'a very large part of the Russian speaking community [refrained] from using this tool' (Trechsel 2007: 44, 6). Trechsel's study demonstrates that more than 62 per cent of voters who elected not to vote online did so because they lacked the necessary facilities (Trechsel 2007: 38). It also points out that 'e-voters do not only differ [from traditional voters] with regard to their socio-demographic and economic profiles, but they also do so [...] with regard to their political preferences' (Trechsel 2007: 49).

To the extent that socio-economic status can be elaborated upon geographical lines, it seems significant that, with the exception of the university city of Tartu, the areas of Estonia which, according to the report of the Estonian National Electoral Committee (2007), demonstrated the highest use of electronic voting among the national turnout were Tallinn itself and its neighbouring counties in the affluent north-west of the country. With barely more than a third of Tallinn's score, the economically depressed county of Ida-Viru on Estonia's eastern border with Russia showed the lowest rate of adoption of the electronic system.

The Organisation for Security and Co-operation in Europe's report on Estonian e-voting went so far as to question 'whether [in future] the Internet should be available as a voting method, or alternatively whether it should be used only on a limited basis or not at all' (OSCE 2007: 2). This lack of enthusiasm has not, however, prevented Estonia from forging ahead with this system: in the European elections of June 2009 58,669 Estonians voted online, while Estonia's municipal elections of October 2009 saw 104,413 people (9.5 per cent of the eligible electorate and 15.7 per cent of those who voted) vote online. By the national election of March 2011 140,846 people (24.3 per cent of those who voted) voted online. It is perhaps notable, however, that Estonia's fellow Baltic nation Lithuania has rejected proposals for online elections (Vaiga 2008).

According to the recommendations on *Legal, Operational and Technical Standards for E-Voting* adopted by the Committee of Ministers of the Council of Europe on 30 September 2004, 'measures shall be taken to ensure that the relevant software and services can be used by all voters.' When Jaak Aab suggests that his government intends (one day) to provide full Internet access to their entire electorate, one might therefore ask whether online voting should be implemented before that day has come. Indeed, that day seems further off than Aab implies: although he suggests that 60 per cent of his compatriots have Internet access, he neglects to mention that, for the vast majority, this access has not included the card-reader necessary for online voting. As Ernsdorff and Berbec (2007: 180) emphasize, these card-readers have not been common for personal use and are costly.

From 1993 to 1997 Tarvi Martens headed Estonia's Data Communication Department, where he championed the introduction of a national electronic identity card. Martens then moved to *Sertifitseerimiskeskus* (SK), the private company responsible for national identity cards, electronic signature systems and card-readers. Since 2003, Martens has worked with the National Electoral Committee to implement an online voting system. He has served as both Chief Operations Officer of SK and the Project Manager for the Estonian e-voting system. Martens has commented that 'the increasing number of e-voters is encouraging. I see two main factors behind it: an increasing awareness of ID-card electronic usage, and a steady, trustworthy and transparent e-voting system which produces increasing confidence in users.'

Despite Martens's confidence in the system, he has admitted that 'we do not possess exact data about number of card-readers installed among ID-card holders.' This tallies with the official Estonian government report on *Internet Voting at the Elections of Local Government Councils on October 2005* (Madise et al. 2006: 8) which reveals that there is no reliable data on the distribution of card-readers in Estonia.

However, Martens's Marketing Manager at SK, Andres Aarma, has been able to shed rather more light on the subject: 'Our estimate for the total number of readers installed is currently around 70,000–80,000. How many of those are in personal use and how many in organizational use we do not know. In the end of 2006 we concluded an agreement with OMNIKEY which won the international tenders for procurement of up to 600,000 readers between 2007 and 2009. The aim is to enable everyone affordable access to the key public infrastructure.'

It may therefore be understood that Estonia initiated an online voting system at a time when nobody knew how many people had access to the technology which would allow them to vote from home – although it was probably not many more than the 3.2 per cent of the electorate who actually used the system. But at least the private company which recommended the system in the first place has now secured a deal to sell the technology necessary to make it work. In this context, there is perhaps something chilling about the final words of a presentation given by Tarvi Martens (2007) on

Estonian e-voting: 'There's no way back.' Processes of electronic democracy, once initiated, are not easy to reverse. This is a mode of democracy in which, paradoxically, it seems we have no choice.

In October 2007 it was reported that Estonia planned to introduce an e-voting mechanism involving the use of mobile telephones (Alas 2007b). Despite initial hopes that direct mobile telephone voting would be implemented in time for the municipal elections of 2009, in December 2008 the Estonian parliament ratified plans to implement this system for the first time at the parliamentary elections of 2011. The system – again backed by technology introduced by SK – required that voters added a new authentication chip to their mobile phones. It should also be pointed out that the Estonian government noted that although it was possible to use a mobile phone *to identify oneself for e-voting*, voters still actually needed access to a computer for the voting procedure itself.

The State of Things to Come?

Where then might this example lead us? Certainly not towards an undilutedly cyberoptimistic perspective. The Estonian example demonstrates, if nothing else, that technological developments do not necessarily result in greater levels of participatory citizenship, democratic accountability or social justice. Indeed, any attempts to deploy these technologies to advance democratic values have been undermined (or indeed led) both by the influence of commercial and political interests and by the failure of the nation's e-visionaries to see that, rather than solving the country's social problems, the imposition of new technologies upon relatively youthful processes of government and democracy may in fact exacerbate those problems.

Even the founding father of the Estonia's Tiger Leap Initiative, Linnar Viik, has expressed doubt over these experiments in electronic democracy. In 1997 Viik returned to Estonia from his studies in Finland bursting with ideas to promote his country's electronic development, and became the government adviser who changed the face of Tallinn's economy. More recently Viik has argued that 'e-democracy doesn't have a real impact on the democratic process. Democracy in Estonia is like a small child. I can compare it to my five-year-old son. He can talk, he knows some manners, he knows how to pee – but he's still learning. This technology is just a tool.'

Yet Viik believes that traditional modes of parliamentary democracy are now obsolete: 'I don't so much believe in representative democracy. I believe in participatory democracy.' Viik's distrust of traditional political structures and processes and his hope that new media technologies will foster democratic participation reflect the view of such studies as those of Coleman and Spiller (2003) and Coleman and Blumler (2009), a view that new media technologies have a crucial role to play in any strategy which hopes to restore public trust and participation in democratic processes. However, for as long as such participation remains limited by educational, demographic and socio-economic conditions, this mode of democracy will continue to be dogged by the concern that it has become the willing instrument of entrenched power structures and privileged interest groups, and that it offers only the illusion of participation in place of real democratic empowerment.

This illusion of empowerment recalls that which Henrik Bang (2010: 261) describes as a process which seeks at once to empower and to *domesticate*. This mode of governance, writes Bang (2010: 247), 'aims at empowering as many people as possible; not primarily for their own sake but to get them to help their organisations to get the minimal wholeness, coherence and effectiveness that they need.' It attempts to 'gain its autonomy as a mode of rational rule via its own specialized semantic, which [...] has the consequence of ignoring those everyday narratives and forms of life that do not possess this kind [...] of specialisation' (Bang 2010: 250). Such empowerment is thus empowering only for those willing and able to fit its model.

In January 2003 Donald Rumsfeld famously suggested that the post-Communist nations of central and eastern Europe epitomized what he called the 'New Europe'. If indeed they represent the latest stage of European development, then perhaps the ways in which Estonia, ostensibly the most post-industrialized of these eastern European states, has adopted and adapted these new media technologies may also reveal something of the economic and political future of European civilization as a whole. Whether this future is revealed as a techno-utopia or a cyberdystopia remains, of course, to be seen. What one might usefully suggest is that neither perspective is entirely convincing – and that in order to move as best we can towards the former state we must remain constantly aware of the dangers of the latter position – that, rather than optimism or pessimism, a continuing critical scepticism might best inform the ways in which we negotiate our progress towards a more relevant, effective and sustainable mode of participatory democracy.

CHAPTER 3

War Games

We have been experiencing, for half a century, a conflation of material history and its electronic mediation, and this phenomenon is perhaps at its most remarkable in the conduct and representation of military conflict.

Jean Baudrillard (1988: 49) wrote of Vietnam as a television war – but Vietnam also of course eventually became a cinematic war, a war primarily recalled in the popular imagination by such films as *The Deer Hunter*, *Apocalypse Now*, *Platoon* and *Full Metal Jacket*. Another postmodern conflict, Operation Restore Hope, America's vain attempt to bring order to Somalia in 1992–93, also began as an event staged for the TV cameras (even to the extent that the Pentagon are said to have consulted CNN on the scheduling of the U.S. landings in Mogadishu), and ended up as a film by Ridley Scott: a five-month military debacle immortalized as *Black Hawk Down* (2001).

The BBC's World Affairs Editor John Simpson's declaration of his personal liberation of Kabul in November 2001 and Donald Rumsfeld's announcement in February 2006 that newsrooms had become crucial battlefields in the War on Terror are two well-known examples of the convergence of media and military perspectives. As Baudrillard (2005: 77) wrote, 'media and images are part of the Integral Reality of war.' Ronald Reagan's abortive *Star Wars* programme stands as a landmark moment in this process, and that Hollywood President's Tinseltown apocalypse was to be echoed in George W. Bush's cowboy diplomacy and in his administration's use of popular filmmakers as strategic imagineers. The influence of Hollywood culture on the conduct of contemporary military activity was underlined by the claim (made by the BBC's 2011 *The Secret War on Terror*) that Guantanamo Bay interrogators aped torture techniques from scenes in the American TV thriller series *24*. One of the most extraordinary convergences between the military and entertainment industries was the news of January 2011 that China's military had used shots from *Top Gun* in official footage of their new fighter plane. A similar confusion of material military reality and entertainment fantasy took place later in 2011 when Britain's ITV broadcast footage purportedly of IRA military training that turned out to be from a video game. As Jeffries (2011) reported, Marek Spanel, chief executive of the game's developer Bohemia Interactive Studio, said that he considered this 'a bizarre appreciation of the level of realism' incorporated into the game. (Such confusion is hardly unique to this incident. The following May, for example, a BBC news report on the United Nations Security Council mistakenly showed the logo for the United Nations Space Command from the computer game *Halo*.)

The overlap between the media and military power is increasingly overt. The jingoistic and war-mongering stances of much of the British and American tabloid press (and, in the United States, of certain TV news networks, most notably the flag-waving Fox – an organization which boasted much-publicized connections with the Bush family) recall Orson Welles's immortal exhortation to a journalist in *Citizen Kane* (1941) – paraphrasing what has been (possibly apocryphally) reported as a line from the real-life newspaper tycoon William Randolph Hearst's in 1897: 'You provide the prose-poems. I'll provide the war.' This is not quite the situation suggested by the James Bond film, *Tomorrow Never Dies* (1997), in which Jonathan Pryce's media mogul attempts to start a war between Britain and China in order to boost newspaper sales and penetrate new satellite broadcasting markets; and yet we are growing ever more acclimatized to that absurdity. (See also Watt 2012.)

Slavoj Žižek (2002: 15) has proposed that even the events of 9/11 itself may be seen as representing the physical manifestation of globalized media product – a Hollywood catastrophe movie: a further blurring of material reality and popular entertainment – in the words of Tumber and Webster (2006: 4) a *media event*. Within five years, the events of 11 September 2001 had already lent themselves to a number of film adaptations including Oliver Stone's *World Trade Center* (2006) and Paul Greengrass's *United 93*

(2006) – and had even prompted a comic book version, Sid Jacobson and Ernie Colon's *The 9/11 Report: A Graphic Adaptation* (2006), a video game called *9–11 Survivor* (2003), and, in 2011, a colouring book for children entitled *We Shall Never Forget 9/11: The Kids' Book of Freedom*. In an article published in *The Times* on 12 September 2001, Michael Gove had written that that 'the scenario of a Tom Clancy thriller or Spielberg blockbuster was now unfolding live on the world's television screens.' The events of that day were especially reminiscent of one particular Tom Clancy blockbuster, a novel entitled *Debt of Honour*, a book which climaxes with a terrorist crashing a civilian airliner into Washington. (It seems no coincidence that CNN chose to interview Tom Clancy during its coverage of the attacks on the WTC.) More extraordinarily, six months prior to the catastrophe of 2001, the debut episode of the Fox TV science fiction series *The Lone Gunmen* had depicted an American government attempt to simulate a terrorist outrage by crashing a hijacked airliner into the World Trade Center itself.

Once more, the so-called and self-styled enemies of the West are influenced and inspired by, and inscribed within, its pandemic media: Stalin was said to be a great fan of American gangster films; Saddam Hussein reputedly enjoyed Scottish football as he ate his Jaffa Cakes in his prison cell but did not live to see his 2000 novel *Zabibah and the King* adapted into a 2012 satire by filmmaker Sacha Baron Cohen; Kim Jong-il was reported to be an aficionado of Hollywood cinema (he once boasted that he owned 'all the Academy Award movies'); indeed, according to his former lover Kola Boof's *Diary of a Lost Girl* (2006), Osama bin Laden himself was a big fan of *Miami Vice*, *The Wonder Years*, Whitney Houston and *Playboy*. As Slavoj Žižek (2008: 73) has suggested, 'the fundamentalists are already like us [...] secretly, they have internalized our standards and measure themselves by them.'

Years before 9/11, Jean Baudrillard (1994: 21) had written that 'all the holdups, airline hijackings, etc. [...] are already inscribed in the decoding and orchestration rituals of the media.' Or, as Slavoj Žižek (2002: 146) has put it: 'Jihad is already McJihad'.

Gameworld

Material history ebbs away, to be replaced by the virtualizing culture of the mass media. This sense of the slippage of the material beneath the mediated is famously elaborated in Jean Baudrillard's *The Gulf War Did Not Take Place*, in which the French sociologist suggests that the Gulf War may be read not as a piece of real, material history but as a media event, a pseudo-event performed for the media. It is not perhaps insignificant that the Gulf War's emotionally sterile images of bombings – images captured by cameras mounted on warplanes, pictures which saturated the television coverage of that conflict – not only exposed a blurring of material-historical and electronically mediated perspectives, but also conspicuously translated acts of mass destruction into the visual idiom of the video game.

Andrew Darley (2000: 31) has argued that the primary benchmark of success in digital game design is the game's graphical verisimilitude, its representation's approximation to external reality. Yet one is tempted to suggest that Darley's argument might be inverted: that the verisimilitude of material reality may now conversely be judged by its approximation to the virtual world. Media texts do not merely reflect reality; as John Fiske (1987: 21) suggests, they construct it. The photographic realism at which the digital game has aimed is – as Bolter and Grusin (2000: 55) point out – precisely that: a realism whose model is not material reality itself so much as the visual perspectives of photography and cinema. And, if the visual realism of the video game is defined by another medium (if it sees – because we have all come to see – the visuality of photography and cinema, with its edits, close-ups and lens flare, as the benchmark of the real), then why should we not see the graphic idiom of the computer game as a new standard for the representation and perception of material reality? Haven't Second Life and *Half-Life* in this way come to represent for many people a first and a full life: a primary and comprehensive mode of existence against which we may now measure the verisimilitude of material reality?

David Nieborg (2006) proposes that – although war games may be founded upon material reality – it is the case that current design limitations prevent the full simulation of the battlefield experience in First Person Shooter games. Again, could we not invert this argument: could we not see that in what Jameson (1991: 48) calls the *society of the simulacrum* the digital game has become a crucial yardstick for the real? If, as Katherine Hayles (2000: 69) suggests, the citizens of late postmodernity live increasingly virtualized lives, then could it not be expected that at the extremes of the real – in a time, for example, of totalizing informational war, a war against an abstract concept, the most absurd of all possible wars – we might begin to witness a situation in which, in the words of Geoff King and Tanya Krzywinska (2006: 200), 'the distinction between reality and simulation might occasionally appear to blur, like something out of the pages of Jean Baudrillard'?

It is not just that the virtual and the non-virtual are becoming indistinguishable; what is significant is that the non-virtual is increasingly subordinated to the virtual. This has been going on for quite some time. In, for example, a 2002 essay on *Counter-Strike*, Wright, Boria and Breidenbach tellingly refer to 'the non-virtual world'. This phrase signals the prioritization of the virtual: the virtual is no longer the 'non-real'; the virtual is not defined by its relation to the real – the real is defined by its relation to the virtual; the real is now merely the 'non-virtual', a category of secondary significance. The digital game, as today's most engaging mode of popular virtuality, comes therefore to represent a primary version of reality. Apparent improvements in games graphics (the narrowing of the gap between representation and reality) are not merely a result of the virtual having become more real: they may also result from the real having become more virtual.

Jane McGonigal (2008) has announced that lessons learned from the development of digital games could usefully be applied to the material world – 'not [to] make our games more realistic and lifelike, but [to] make our real life more game like.' It may in fact be that McGonigal's ambition is already being realized by the processes of cultural evolution. In March 2010 the World Bank launched a digital game designed by McGonigal: entitled *EVOKE*, it promised 'a ten-week crash course in changing the world' by offering to 'empower people all over the world [...] to come up with creative solutions to urgent social problems.' The gameworld is gaining existential primacy: the gameworld, for many, already *is* the world.

The Dogs of War

One is reminded in this connection of that avatar of virtual existence, 'FPS Doug', an obsessive digital gamer in the cult video series *Pure Pwnage*. *Pure Pwnage* was a series of low-budget mockumentaries which fictionalized the lives of the digital games players of Toronto. The series ran online for 18 episodes over two seasons from 2004 until 2008, before moving to Canada's Showcase cable television channel for a further eight episodes in 2010. The series's title (pronounced 'pure ownage') refers in gaming slang to a player's mastery of their game.

In the fifth episode of the series we see FPS (First Person Shooter) Doug taking a knife with him when he goes jogging because he believes that in the non-virtual world, as in the virtual, one can run faster with a knife. Doug's primary reality is the virtual: as he suggests in the same episode: 'Sometimes I think maybe I want to join the army. I mean it's basically like FPS, except better graphics.' But is it not increasingly the case that, insofar as we view the material world through prevalent visual paradigms, the reality of military experience resembles the video game, albeit with somewhat *less* convincing graphics?

Pure Pwnage's lead character, games addict Jeremy similarly sees the world through the perspective of the video game. In the premiere episode he asserts the irrelevance (indeed the existential inferiority) of the non-virtual world: 'normal people wake up in the morning and they watch CNN. I just don't like that because that's kind of fake. Where's the RPGs [role-playing games]? – it's all about terrorists and Bush and stuff.' For Jeremy, the world of war and politics has become less real than the realm of the role-playing game. In the series's second episode, for example, when he enters a bar in a somewhat futile attempt to pick up women, he supposes that the bar is 'kind of like a level that you play on.' When in the sixth episode Jeremy discovers the game *World of Warcraft*, he announces that he has 'a new place to live' – that he is 'going to live in Azeroth.' Instead of taking a break from the game, he says he can 'stay at the inn, get some rest' within the game. When asked what he will eat, he points out that 'there's food and drink in the game.' His desire is for his real life to be subsumed to the game, as if he might perform the fantasy of total immersion played out in the *Tron* films (1982, 2010). By the following episode, Jeremy has entered a digital psychosis, as his consciousness flashes between lifeworld and gameworld and he is unable to distinguish between the two.

At the end of this episode Jeremy appears to have recovered; yet this marks something of a turning point in the series's narrative. From this point on, the series becomes increasingly game-like itself: it is not that Jeremy has returned to material reality so much that this blurring of the material and the virtual has shifted the audience into an increasingly surreal or hyperreal world. When, for example, in the eighth episode, Doug is shot in an FPS game, he collapses in the real world. By this episode, although Jeremy's doctors have banned him from playing *World of Warcraft*, he is trying to establish his own 'gamer army' in the real world. As the series progresses, Jeremy's battles in the physical world increasing echo the fantastical polarizations of the gamewold as he assembles his allies and confronts his nemeses. His material life itself takes on the characteristics of a video game – to the extent that by the eleventh episode we discover that video game laser guns can work in real life.

One of the recurrent and defining characteristics of the superhero genre is that, when not assuming the mantle of their superheroic alter ego, the protagonist leads a normal human life – a life which grounds them in everyday morality – while the genre's villains tend to be caught within their surreal personae (rarely does one see the Joker, for example, bereft of his make-up and attending to his day job). In this way, these fables may act as warnings against an addiction to an alternative selfhood – when, as Jim Carrey discovers in *The Mask* (1984), or as Robert Louis Stevenson's Henry Jekyll and H.G. Wells's Invisible Man had found a century earlier, the metahuman self corrupts and overwhelms the original self. This danger of irreversible transformation is clearly addressed at those moments when superheroes choose to reject their assumed personae in order to reassert their humanity – as when, for example, Michael Keaton rips off his mask towards the end of Tim Burton's *Batman Returns* (1992). Yet the humanly impotent self is for the most part perhaps inevitably subsumed to the superhumanly powerful self, even when (as for the digital gamer) that superhuman power is transitory and illusory. If all subjectivity is performative, and if that performativity is determined by external parameters, then an imaginary super-agency is clearly more attractive than the reality of mortal struggle and near-powerlessness. Faced with an aggrandizing virtual existence whose swift seamlessness offers little room or reason for critical self-reflection, the gamer may thus all too easily succumb to the temptations of fleshlessness – the real desert of the real (that is, the desertion of the real), a fate which Joe Pantoliano's character Cypher seeks to embrace in the Wachowski brothers' *Matrix Reloaded* (2003). As *Pure Pwnage*'s Jeremy announces in the series's eleventh episode, 'in the future we're going to be living in tombs – we're just going to have stuff hooked up to our brains – and we're basically just going to sit round playing games all day long.' The Wachowski brothers' dystopia is Jeremy's utopia.

Jeremy can no longer distinguish between virtual reality and real life, and no longer wants to: 'real life' is merely 'RL' to him, an inferior alternative to VR: 'RL is a game.' A fantasy sequence at the end of the series's twelfth episode, which concludes its first season (and to which the series returns in the first episode of its second season), shows both these realities collide, as RL is apocalyptically overtaken by VR. As one games enthusiast says in the second episode of the third season (explaining why she has just taken revenge for a slight in her gameplay by shooting a real gun in real life at a man who had offended her): 'just because something's virtual doesn't mean it's not real.'

At the end of the series's second season, Jeremy reveals the name of the video game he has himself been planning to design – 'the best name ever' (though he mistakes its meaning) – *Vasectomy*. The irony is that what Jeremy sees as a term which symbolizes creative energy, agency and power is, of course, synonymous with a self-imposed infertility. This again is that barrenness, that desert(ion) of the real sustained by a mirage of empowerment in which the digital escapist or fantasist has elected to engage their sense of self.

Pure Pwnage producer Davin Lengyel has pointed out that the series's game-obsessed characters were all based on real people – either exaggerated versions of the actors and writers themselves or the people they play games against. He has suggested that the popularity of competitive games is the way in which they offer validation to their players: 'you can say you are better than someone else.' Lengyel has gone on to express some concern as to how much these games may influence their players:

> My experience is that there is a bit of crossover, and I think people need to be aware of it. I've personally been influenced in my personal behaviour by games I play. I noticed this in particular with *Grand Theft Auto 4*. Everything in that game revolves around the idea that you are going to be a violent person. I'd play six or seven hours a day, and when I walked outside and I'd see a car of the same model as in the game, I'd have a desire to steal that car. If you practise a behaviour all day, it becomes natural to you. It's quite easy for people to take on learned behaviours. If I play this game for months, I'm going to be influenced by this game. You do spend a lot of time with your games – it's what you think of when you're at work – you want to get home to play your games.

Lengyel is particularly concerned by the way in which such games as *America's Army* may be used literally to recruit their players:

> It's no surprise there are games being put out by the military. The U.S. military has developed training simulations which are kick-ass video games. It's a natural fit. But to use video games as a means for recruiting for the armed forces for me is ludicrous. They're using a medium of entertainment to create a positive brand for the U.S. military and to desensitize people to the dangers of combat. It's very clever, slightly devious and a little frightening. There's a new version of a flight simulator, sponsored by the U.S. Air Force, all their models provided by the Air Force – their slogan is 'This time it's real.' It makes me really sad when I hear soldiers saying it was very much like a video game when they were in combat. The learned behaviour is you can get shot and still win. It scares the hell out of me that this might be a learned behaviour that people might take into a combat situation. The soldiers in combat have to remind themselves what they're doing and that it's actually real – and that the person they've just shot was a real person.

Lengyel also notes that the confusion of real-life conflict with the digital game is taking place in the deployment of such military technology as remotely piloted aircraft, drones whose missions in the Middle East may be piloted by service personnel in the United States: 'People who pilot drones – they're in the States – they shot and kill people and fly the drone with a joystick – it's not dissimilar to a video game.' Game/life boundaries are clearly being blurred. Citing an interview with one such drone pilot,

Lengyel has observed that the pilot had commented that the emotional impact and immediacy of this mode of warfare derives primarily from its generic and morphological familiarity, that it looks and feels just like a video game: 'killing a human is so easy and so familiar – like a video game.' The game has become the yardstick of reality (that is, it seems to be perceived, paradoxically, as the ostensibly originary reality); in Baudrillard's terms the virtual map has not only superseded but appears now to pre-empt the physical territory. We do not feel the emotional and moral significance of another's life and death in that they are of the same biological species, the same flesh and blood, as us – but only insofar as they are incarnate avatars of beings from a digital realm.

In the last episode of the *Pure Pwnage* TV series FPS Doug's gaming reality finally overtakes his material existence, as the devotee of the FPS game *America's Army*, recruited at a gaming convention, steps up to enrol in the military: 'what kind of adventure is this – RPG style?' It is difficult to see this as an act of empowerment. 'I'm not an idiot,' he says. 'Sure, I'll die a few times but there's no way I'll be dying as much as them.'

In March 2003 the UK's *Sun* newspaper had used a gaming metaphor when it ran on its front page a photo of a British soldier brandishing his gun beneath the headline: 'Game over: Blair tells Saddam his time is up.' Three days later the invasion of Iraq began.

The distinctions between the video game of the war and the war of the video game have significantly blurred. In 2002 the United States military launched a First Person Shooter digital game entitled *America's Army*. The game was first intended as a training tool, but was swiftly repackaged and made available at no cost online for propaganda and recruitment purposes. In December 2011 the game's website boasted that it had served more than 12 million registered users.

In 2008 the website for *America's Army* promoted its latest edition, *America's Army: True Soldiers*, as 'the only game based on the experiences of real U.S. Army soldiers.' This version of the game announced that it had been 'created by soldiers, developed by gamers, tested by heroes.' It is notable that this promotional copy recognized no final distinction between soldiers and gamers: both are *heroes* – neither of them *play* the game; they both *test* it. This testing serves multiple functions: the reality of the simulation is tested by soldiers and gamers alike to temper and strengthen its military value as a training tool, while that simulated reality thereby becomes the dominant version of perceived reality, and thus works as a tool both for propaganda and for (ideological and actual) military recruitment. When *America's Army 3* was launched in June 2009, it continued to stress its proximity to the real: 'Your journey will begin with training so realistic you'll swear you're actually there. That's because the training in *AA3* was created by the U.S. Army and is based on real Army training. In *America's Army 3*, characters are more authentic than characters in any other video game. Every detail about *AA3* weapons has been verified by Army Subject Matter Experts. From the way the weapons look to how they are deployed and how they sound, the level of realism is unparalleled.' The website was also updated to include profiles of real soldiers serving in the U.S. military awarded for bravery; their integration into the paratext of the game again emphasizes not only the verisimilitude but in effect the reality of the virtual experience: 'The opportunity to meet these Real Heroes is one of the many ways that *America's Army 3* is a game like no other.'

Zhan Li (2004: 137) has suggested that *America's Army* represents an ambiguous space caught between the political, the military, the commercial and the material. This blurring of traditional generic, ontological and epistemological boundaries is perhaps best evidenced in *America's Army*'s touring recruitment circus, a disconcertingly physical roadshow – the chance to see some 'real' U.S. military hardware, alongside videos of serving soldiers and, of course, at the forefront of all this (as the prioritized mode of reality), the gaming experience itself. According to its website, this Virtual Army Experience (VAE), as it calls itself, 'provides participants with a virtual test drive of the Army, with a focus on operations in the Global War on Terrorism.' It announces that its participants 'enter the mission simulator area where they execute a simulated operation in the War on Terrorism.' There is something disappointingly incongruous in the physical aspect of the VAE's recruitment roadshow; its physical presence seems to represent a poor copy of the digital game itself. This phenomenon is hardly unique to *America's Army*: when, for example, *World of Warcraft* fans don their home-made costumes and enact gameplay performances at their conventions and other such gatherings, are not these materializations of their gaming somehow less realistic, and therefore (to them) less real, than the virtual world they attempt to emulate?

Re-establishing the primacy of the virtual over the material, the *America's Army* website offers a virtual tour of the VAE – a simulation of a simulation, one which leads the eye through a computer-graphic reconstruction of a room full of computers, their screens displaying scenes from the original game. This virtual tour does not allow the user any navigational control: like a gameplay video, it leads its impotent viewer through its environment – towards the inevitable end of interpellation and recruitment: to sign up to the U.S. military or at least to its ideological perspective. The VAE's virtual tour represents a meta-simulacrum – an electronic simulation (a virtual tour) of a physical simulation (the VAE touring event) of an electronic simulation (*America's Army* – the game) of a military reality which is increasingly virtualized. Bolter and Grusin (2000: 216) – in the context of a similar instance in which computers appear within the digital virtualscape – ask why we see such computers in virtual worlds: what would virtual worlds need them for? One might suppose that the presence of the computer within the virtualscape emphasizes at once the reality of that virtuality (reality has computers in it; indeed much of our reality is defined by their presence; so a virtual world seems more real if it has computers in it) and the ubiquity – in effect, the definitive significance – of the computer within *any* version of reality. As such, the presence of the devices of the virtual within these virtual imaginings not only blurs the distinction between the virtual and the material but also simultaneously prioritizes the virtual (or the digital) above analogue reality.

This process is clearly seen in the conduct of war itself. Warfare has, after all, begun to adopt the characteristics of the digital game. King and Krzywinska (2006: 199) point out that material warfare is increasingly mediated by such devices as head-mounted displays offering visuals highly reminiscent of video games. The distance between the soldier and the gamer is blurred, an effect of technological developments whose dissemination has been accelerated and intensified by the War on Terror. As Tumber and Webster (2006: 33–34) stress, game theory and digital simulations are essential elements in the conduct of contemporary warfare. While

David Nieborg (2006) notes that the same military simulations are used by both soldiers and gamers, Edward Castronova's analysis (2005: 234) goes somewhat further when he suggests that 'the emergence of open-source military game-building tools has effectively turned the entire world into a giant military research lab.'

King and Krzywinska (2006: 199) suggest that gameplay may be used to train players in the techniques of realworld warfare. This facility is not only, of course, the province of the United States and its political, military and ideological allies. Such video games as *Under Ash* (2001), *Special Force* (2003) and *Under Siege* (2007) have promoted (and have been used to train) anti-Israeli paramilitary groups in the Middle East; while even such overwhelmingly neoconservative toys as *Counter-Strike* (1999) might also offer, as Castronova (2005: 231) suggests, a convenient tool for training terrorists.

This training is as ideological as it is strategic. The First Person Shooter game foregrounds the gun, both visually and linguistically. The first person (the ego) becomes identical to the 'shooter' – both the person who shoots and the gun itself, and indeed also the imagined lens of visualization. Within that one word – *shooter* – the distinction between subject and weapon dissolves. The player is translated into an organ of war shooting forth its deadly seed to inseminate a new world order. *America's Army* has promoted itself beneath the slogan 'empower yourself – defend freedom'; yet what it offers is disempowerment, a loss of freedom, a loss of self and even the player's actual death (it recruits you; it can get you killed). It therefore seems significant that, as Lars Konzack (2009: 39) points out, the multi-player mode of *America's Army* deploys an extra level of illusion in order to sustain its players' ideological assimilation: each player sees themselves as the U.S. soldier and the other as the terrorist – though each is always also the terrorist from the other's perspective.

A 2010 recruitment commercial for Britain's Royal Navy began by emphasizing the meaningless routine of our daily lives: 'You're born, you cry ...' Like *America's Army* the Royal Navy then offered itself as a release from the trap of mundane existence: its alternative was 'a life without limits.' It is, once more, an illusion of empowerment that recruits us, that seduces us into dropping our resistance to our assimilation and disempowerment.

In 2011 a TV commercial advertising life in the British armed forces suggested it represented a more valid form of existence than surfing the net all day – before giving a website address for more information. Military experience is proffered with all the seductive promises of these escapist electronic fantasies.

The Illusion of Agency

Those familiar with the video-sharing website YouTube may be aware of the practice by which digital gamers have edited together clips of – or merely recorded extensive sequences of – their gameplay, added music or voiceovers and credits to it, and posted it online as a so-called 'movie'. This practice has (thus far) reached its most extreme expression in *Red vs. Blue* (2003–2007), a 100-part narrative reconfiguration of the video game *Halo* (2001). The experience of watching one of these gameplay films is nauseating: not only in a physical sense (like being a passenger in a drunk driver's car) but also in existential terms. This is the Sartrean nausea which accompanies the realization of one's own lack of control, the revelation that one's feeling of self-determination was only ever an illusion – that the experience of playing the game and of watching the game being played are, in the end, the same. The amateur gameplay video therein exposes the possibility that the digital game's defining sense of a player's agency may be illusory, the realization that the illusion which re-envisages the immutable, impersonal edifice of the game as an extension of the gamer's own subjectivity – as if the player could somehow reconfigure the programmatic structure of the game, could escape its pre-programmed linearity (which is a multilinearity, but is still a linearity, and a finite one at that) – is the precise opposite of what is in effect taking place. In watching gameplay with the detachment of an uninvolved audience, the game's inescapable parameters and its pre-programming become visible.

Lars Schmeink (2008) has referred to the mere *sense* of agency afforded by the video game; for, as Tanya Krzywinska (2008) has suggested, 'you are promised some kind of agency, but your agency is taken away from you.' This sense of agency is always, she adds, a fictive agency. James Newman (22002) has argued that videogames are not interactive and Dominic Arsenault and Bernard Perron (2009: 119–120) have similarly challenged the popular notion that the video game is a predominantly interactive medium. They argue that in fact players are not active but reactive – that players respond to pre-programmed structures within the game, structures designed to predict and react to the gamers' responses. The illusion of interactivity sponsors a sense of agency – but this agency has been externally predetermined or pre-designed.

This illusion of agency is central not only to the pleasure of the video game but also to its purposes (to its uses, then, as well as to its gratifications). As Klimmt and Hartmann (2006: 138) write:

> Players experience themselves as causal agents within the game environment [...] computer game play [...] produces many individual experiences of efficacy and causal agency [...] Most games allow players to modify the game world substantially through only a few inputs. For example, in a combat game, players often need only a few mouse clicks to fire a powerful weapon and cause spectacular destruction. The ability to cause such significant change in the game environment supports the perception of effectance, as players regard themselves as the most important (if not the only) causal agent in the environment.

The game sponsors an illusion of agency which places the player at the centre of its universe – which thereby for the player becomes *the* universe. This illusion is not immediately recognisable as such; from the player's perspective the game's interpellating devices remain invisible.

The game's self-proclaimed interactivity is not a case of co-authorship: the gamer is funnelled through a limited and limiting series of preset positions. The simulacrum of the digital game constructs and delineates its citizen-user-consumers as avatars of its own subjectivity. The gaming subject is *interpellated*, in Marxist theorist Louis Althusser's sense – is hailed or recognized as the central character in their sphere of existence – and, as this process is never without an ideological destination, the subject is posited within (or, in the case of *America's Army*, literally *recruited* to) a new reality. 'Ideology,' writes Althusser (2006: 118), '"recruits" subjects among the individuals (it recruits them all), or "transforms" individuals into subjects (it transforms them all).'

This represents a process of ontological and ideological transformation. To what extent then does the gamer become subsumed to the subjectivity of the game? Although Bolter and Grusin (2000: 253) have suggested that the visitor to virtual reality remains aware of the differences between the virtual and material worlds, they have nevertheless supposed that virtual reality changes our notions of selfhood. More recently King and Krzywinska (2006: 198) have asked:

> Are players, really, interpellated to any significant extent into the *particular kinds* of subjectivities offered by the in-game diegetic universe? [...] Plenty of markers exist that clearly announce the large gulf that exists between playing a game [...] and engaging in anything like the equivalent action in the real world. But there are, also, certain homologies. How far these come into play depends on a number of factors, including the [...] forms of realism [...] which can shape the extent to which the game experience approximates that of the real world.

Yet, as has already been suggested, the extent to which the world of the game approximates a prevalent notion of the real world may matter rather less than the degree to which the subsidiary, material world resembles the hegemonic gameworld. As the reality of the game becomes the dominant mode of being, King and Krzywinska's gamer is increasingly assimilated within the gameworld's subjectivity.

In his anthropological survey of Second Life, Boellstorff (2008: 120) observes the formation of distinct identities in virtual worlds. One might argue that the senses of safety and of empowerment generated by this virtual environment give the user an impression of autonomy that fails to register – and therefore to resist – the constraints and influences placed upon it, and that therefore the blurring of offline and online identities through what Boellstorff (2008: 121) calls a 'permeable border' between selves might foster modes of offline being inconsistent with the traditional paradigms of the material world. Boellstorff (2008: 122) comments that a number of the Second Life residents within his study 'spoke of their virtual-world self as "closer" to their "real" self than their actual-world self.' If it is the case, as this suggests, that the virtual self becomes the user's primary benchmark of subjectivity, there might appear to be something increasingly problematic in this identity seepage between these two modes of being.

Edward Castronova (2005: 45) has proposed that the gaming avatar is no more than an extension of the player's body into a new kind of space – as though the assumption of a mask or a persona does not transform one's identity. Slavoj Žižek (2008: 83) adopts a rather more ontologically problematic perspective: 'when I construct a "false" image of myself which stands for me in a virtual community in which I participate [...] the emotions I feel and "feign" as part of my onscreen persona are not simply false. Although what I experience as my "true self" does not feel them, they are none the less in a sense "true"'. Gameplay constructs an alternative but real subjectivity, and, insofar as the gamer increasingly experiences the virtual world as her primary reality, then that alternative subjectivity may come to represent the player's dominant sense of self. Indeed, as Žižek (1999) has also suggested, the game both constructs and constrains the self – and therefore this play is far from liberatingly transformative:

> The mystification operative in the perverse 'just gaming' of cyberspace is thus double: not only are the games we are playing in it more serious than we tend to assume (is it not that, in the guise of a fiction, of 'it's just a game', a subject can articulate and stage [...] features of his symbolic identity that he would never be able to admit in his 'real' intersubjective contacts?), but the opposite also holds, i.e. the much celebrated playing with multiple, shifting personas (freely constructed identities) tends to obfuscate (and thus falsely liberate us from) the constraints of social space in which our existence is caught.

David Myers (2009: 48) suggests that 'when we play with self, that self is something other than what it is: an *anti*-self' – an alternative, and possibly overpowering, mode of being. Ian Bogost (2006: 136) also recognizes the tensions between these subjectivities – at the point at which the reality of the game blurs with the material world.

Sébastien Genvo (2009: 135) has suggested that the video game player may be 'engrossed in his game although he knows that after all it is only a game.' The notion of the integrity of identity in the face of cultural or virtual immersion requires, however, the existence of an *a priori* subjectivity – founded upon the romantic notion of an essence of selfhood – or upon the prioritization of material experience as somehow more influential upon the propagation of subjectivity than digitally mediated experience (as though our physical interactions might for some reason mould our identities more forcefully than those hours spent in the virtual space of the electronic media).

There is, of course, no difference between material and virtual experience: it is just that we tend to use the word 'virtual' in depicting forms of experience mediated by more recently evolved technologies. We are defined by performance and play as much as by 'real life' activity – insofar as there is, of course, no difference between these phenomena, except one imposed by culturally, economically and ergonomically determined epistemologies. Jean-Paul Sartre (1969: 59), in *Being and Nothingness*, famously describes the way that a waiter in a café plays at being a waiter: 'all his behaviour seems to us a game.' The waiter is playing at being a waiter: he is playing at being himself. Sartre's point is that it is such play or pretence which defines identity: as existence precedes essence, the parts we play define our subjectivities. If we are all playing parts, then the parts we self-consciously play cannot be differentiated from those roles we unconsciously assume. We are all method actors, and, as Camus (1975: 75) suggests, the roles we create and perform return to create and perform us:

> To what degree the actor benefits from the characters is hard to say. But that is not the important thing. It is merely a matter of knowing how far he identifies himself with those irreplaceable lives. It often happens that he carries them with him, that they somewhat overflow the time and place in which they were born. They accompany the actor, who cannot very readily separate himself from what he has been. Occasionally when reaching for his glass he resumes Hamlet's gesture of raising his cup. No, the distance separating him from the creatures into whom he infuses life is not so great.

The ancient Chinese writer Zhuangzi famously imagined that the man who has dreamt himself a butterfly does not know if he is not a butterfly who is now dreaming himself a man. It is clearly unclear which of our performed or imagined selves is the real one, or indeed whether there is any such thing as a real one. Slavoj Žižek (2008: 83) notes the way in which pretence, performance or play has through the course of human history generated 'real' subjectivities. When immersed in performative activity (as we always are) our suspension of disbelief creates an identity for whom our belief is permanent and absolute.

In his mystery crime novel *Virtually Dead*, Peter May describes the attractions and seductions of the virtual world of Second Life. May himself spent a year researching Second Life, even going so far as to establish his own virtual detective agency therein. The central character of May's novel, Michael Kaplinsky, takes on the persona of crime-fighting Chas Chesnokov in Second Life, a role which impacts upon his own identity, existence and survival in the material world (May 2010: 105): 'Who would he be when he logged out again? Michael or Chas? Or was it possible that, with time, more and more of Chas would return with him to RL?' Indeed Second Life eventually comes to seem to him preferable to real life (May 2010: 142): 'If only he could just be subsumed to the virtual.' He discovers that 'in a world where the reality is virtual, and completely unreal, it is far easier for us to be our real selves' (May 2010: 193); and yet this real self – this increasingly dominant self – is clearly different (although decreasingly distinct) from the original material subject.

But if there is no difference between the ways in which material and digital experience construct subjectivity, should the notion of identity within the virtual realm in any way concern us? What is different, of course, about contemporary digital culture is its globally homogeneous nature, and (through the speed and seamlessness of its operation) the ease with which it disguises its ideological and economic construction. The virtual environment, like any mode of conventional realism, smoothes out the wrinkles in material reality, offering a realm resembling Borges's Tlön, one whose continuity of logic makes more sense (and appears more realistic) than the incoherence of the material world. Its realism offers an immersion in the ultimate escapist fantasy – the fantasy of ontological logic, the fantasy that the reality of experience might *make sense*.

John Fiske (1987: 24) has suggested that conventional realism 'reproduces reality in such a form as to make it understandable. It does this primarily by ensuring that all links and relationships between its elements are clear and logical, that the narrative follows the basic laws of cause and effect,

and that every element is there for the purpose of helping to make sense.'
Material reality of course lacks this seamless continuity: a cosy continuity
which makes things so understandable that we do not make the effort to
understand them. The video game, like any product of populist culture,
embraces its user in that reassuring logic.

Even if she were not lulled into critical complacency by the faultless
logic of the virtual experience, it is also apparent that the digital game's speed
of operation barely allows its user time for such independent reflection. In
their discussion of early film, Adorno and Horkheimer (1979: 127) argue
that the relentless speed of the cinema prevents the possibility of reflective
thought on the part of its audience. The velocity of the virtual world of
the early twenty-first century of course leaves Adorno and Horkheimer's
cinema standing.

The transcultural uniformity and universality of the digital domain
also suggest a paradigm shift in the mediation of identity. For the first
time in human history, the cultural difference which gave that history
its momentum appears to be in the process of being replaced by a single
world view, a ubiquitous and monolithic mode of mediation, representation and perception. It is not the 'virtuality' of digital culture so much as
its globalization which underpins its potential to determine subjectivity.

The homogeneity, seamlessnness, rationality and apparent safeness
of the virtual environment are precisely the factors which may reduce its
users' ability to resist its influences. When we are immersed into the subjectivity of our chosen avatar, we tend not to notice the extent to which
that avatar may have chosen us (insofar as our selection is anticipated
and determined by the avatar's own design), and the impact that it may
thereby have upon us.

But the most powerful hold that the virtual environment maintains
over its users must surely lie in the impression of empowerment which it
offers, an empowerment founded upon the promise of interactivity.

The Illusion of Interactivity

A still influential view of textual interactivity was advanced some decades ago by the cultural theorist Roland Barthes in his elucidation of the *scriptible* text in *S/Z* and in his celebration of 'The Death of the Author'. Barthes argues that traditional texts represent what he calls the *lisible* or readerly: fixed, final and finite products, rather than ongoing processes of interpretative production (Barthes 1974: 5). Their modes of reception offer referenda as passive and empty as the voting practices of *The X Factor*. Barthes's antithesis to this classic readerly text is the writerly or *scriptible* text. Barthes's textual ideal is founded upon the premise that the proper or progressive function of literature is to transform the reader from a passive consumer into an active producer of meaning (Barthes 1974: 4). The writerly text invites, embodies and requires cooperation and co-authorship: it understands that meaning is an act of interpretation rather than of intention or expression. As Barthes (1977: 148) proposes, the intertextual polysemy of the work of art originates where it is destined to end: in the mind not of its author but of its audience.

Sheila Murphy (2009: 197) has reminded us that when Mattel launched its home video game system in the late 1970s, it deployed the striking slogan: 'This is intelligent television.' There are those, however, who question whether smart technologies beget smarter consumers. In contrast to Roland Barthes's textual idealism, there is a school of critical thought which suggests that the political function of popular culture is to dumb us down and that new technologies intensify this process. Noam Chomsky (1989: 14), for example, has famously argued that 'the media are vigilant guardians protecting privilege from the threat of public understanding and participation.'

There is nothing spectacularly new in this idea: in *The Dialectic of Enlightenment* Adorno and Horkheimer (1986: 120–167) complained that cinema's homogeneous processes divested its audiences of the power of critical thought. Bertolt Brecht (1978: 187) meanwhile imagined the users of industrial culture as ideological zombies. Brecht's image all too easily fits the stereotype of the TV addict or the video game junkie. Yet, rather more recently, the likes of John Fiske and Stuart Hall have argued against the absolutism of these hypodermic theories of mass-cultural influence: 'I do not believe that "the people" are "cultural dopes"; they are not a passive, helpless mass incapable of discrimination and thus at the economic, cultural and political mercy of the barons of the industry' (Fiske 1987: 309). Roland Barthes's *scriptibilité* anticipates Stuart Hall's notion that the act of decoding a text may not be equivalent to the process of its encoding – but may encompass negotiation with, or opposition to, the dominant meanings privileged by the position of authorship.

Yet perhaps no texts or media forms are truly negotiable or interactive in themselves. Rather than Hall, Barthes or Fiske's celebrations of the potential of audience co-authorship, it may be that Walter Benjamin's ambivalence offers the most convincing theoretical stance. Walter Benjamin (1992: 234) ambiguously describes the mass media audience as an absent-minded examiner. He proposes that 'a man who concentrates before a work of art is absorbed by it [but] the distracted mass absorb the work of art' (Benjamin, 1992: 232). The former state of immersion permits the survival of an integral subjectivity (not a pre-existent essence but a pre-textual self); the latter process incubates an ideological identity within the passive subject. We remain caught between these positions – between the liberal's free-thinking citizen and the Marxists' dope – or perhaps, rather, we are both (and neither) of these at the same time. We can only be the former when we believe we are the latter; when we believe we are the former, we become the latter.

If those popular texts, technologies and practices which invite audience participation (reality television, competitions and lotteries, phone-ins, teleshopping, electronic governance, citizen journalism, Facebook, Wikipedia and YouTube, online gambling and digital games) in fact offer only an *illusion* of interactivity, then – rather than promoting participation – they may in fact serve entrenched structures of power by sublimating our desires for active, participatory citizenship. It may be argued that the video game's illusion of *scriptibilité* seduces the player into neglecting the modes of critical negotiation which might prevent the states of ideological assimilation envisaged by Adorno and Brecht – that the game's demands for functional reactivity promote an illusion of agency which lulls the player into an interpretative passivity, and which thereby serves to posit its subject within a virtually invisible (and therefore virtually irresistible) ideological mould. This illusion is central to any process of textual interpellation, but the digital game reinforces it with an apparently unprecedented degree of influence. The video game is neither more nor less interactive than any other mode of textuality – yet the video game announces its interactivity more forcefully than perhaps any other media form.

We might therefore add a third category to Roland Barthes's classification of *scriptible* and *lisible* texts: the *faux-scriptible* text which proclaims its openness to interactivity, which gives its user the illusion of meaning, power and active participation, and which, in appearing to satisfy its audience's desire for agency, in fact subsumes and dilutes that desire. This process resembles a kind of textual karaoke: its audiences believe that their participation represents a form of activity, a mode of agency, but they are, in effect (and in consequence), mere puppets of the text. This *faux-scriptible* text is thus significantly more reactionary and compelling than the *lisible*.

Lev Manovich (2001: 56) has suggested that the products of culture have always already been interactive in various ways. Manovich therefore argues not only that the notion that the textual interactivity offered by information technologies is a new phenomenon is itself misleading, but also that this 'myth of interactivity' (as he calls it) does not offer its audience agency but an illusion of agency which is the equivalent of an Althusserian act of interpellation. Manovich (2001: 61) writes:

> Before, we would read a sentence of a story or a line of a poem and think of other lines, images, memories. Now, interactive media asks us to click on a highlighted sentence to go to another sentence. In short, we are asked to follow pre-programmed, objectively existing associations. Put differently, in what can be read as a new updated version of French philosopher Louis Althusser's concept of 'interpellation', we are asked to mistake the structure of somebody's else mind for our own.

It is not that the digital game, for example, is necessarily any less interactive than any other text (and, despite the somewhat naïve and outmoded dichotomy between game and text which some schools of digital game theory have attempted to impose, the game is a text, just as all textuality is play – insofar as text is an intertextual, dialogical, fluid and paradoxical phenomenon); it is that its claims to greater levels of interactivity (and therefore of audience agency) are seductively misleading.

Let's Stop Playing Games

New media have, in their proliferation and domination, attempted to erase the traces of their own material production: in doing so, they defy the possibilities of metatextual reflection and device-baring through which the self-consciousness of Modernist or Postmodernist culture might afford its audience the canny detachment which invites a Barthesian writerliness. Bolter and Grusin (1999: 24) have, for example, written of digital technology's attempts to deny its own mediated character through the promotion of the notion of a transparent or invisible interface. Our immersion in the digital experience allows us to deny or to ignore the fact that the experience is merely digital; or, rather, this denial – this suspension of our awareness and therefore of our disbelief – remains essential to the processes of immersion.

Ernest Adams (2009) has suggested that video games are like Victorian novels: despite their claims of interactivity, they remain, in narratological terms, classic realist constructs which eschew the disruptive and liberating possibilities of a metatextual *scriptibilité*. The enduring presence of the fourth wall sequesters the player against a self-consciousness which might foster a reassertion of the critical self. The mass-market digital game's refusal to bare its aesthetic devices thereby allows its ideological mechanisms to go unchallenged and unseen. Though the popular video game may allow an awareness of its artificiality (it may let you know it cannot really kill you – not at least within its own primary conditions of reception), it cannot afford to admit to any challenge to the coherence of its artifice, because that would be to undermine its interpellating structure of realism. Its conventions – its rules – can be asserted but cannot be questioned. Playing a commercial video game is therefore rather more like walking – or boarding a rollercoaster or an aeroplane – than it is like reading *Finnegans Wake*: the more one thinks about it (the more one questions the natural cogency of its conventions of function and production), the less possible it becomes.

Adams (2009) has gone on to suggest that in multiplayer games 'the author ceases to be an author' – and that this role is therefore devolved to the player. There are convincing arguments that in multiplayer (and other) games the reader may become equivalent to an author (albeit for the most part an unread one) – but this is not the same as Roland Barthes's ideal of the reader-as-producer. Barthes dreamt of the death of the author; yet, by contrast, the multiplayer experience does not dissolve authorial authority, it disseminates it. The multiplayer game does not promote the interactivity of co-productive and dialogical readership so much as it proliferates authorship as a set of parallel but solipsistic or monologistic experiences.

There is, of course, the hope that the digital game, a media form still in its infancy, will mature into the modes of complexity and ambiguity, of *scriptibilité*, which cultural theorists have witnessed in the triumphs of Modernist and Postmodernist literature, painting and cinema. Ernest Adams has supposed that this may eventually happen, while David Hayward (2008) has suggested that, despite the intellectual conservatism of the commercial games industry, truly inventive digital games may one day be recognized by wider audiences. Alexey Pajitnov's *Tetris* (1984) is of course the benchmark for intellectual innovation in games production; yet its refusal to inscribe its potential for interactivity within a narrative textuality and material context may suggest to some the eventual incompatibility of the ludological and narratological frameworks which have underpinned academic theories surrounding digital games. The same reductive logic might eventually lead one to identify Microsoft Word as the world's most popular and interactive digital game, and there may be something revealingly resonant in that absurdity.

It may, however, be that such games as independent designer Gonzalo Frasca's *September 12* (2003) and *Madrid* (2004), although as structurally simple as *Tetris*, suggest a route towards the textual sophistication of the digital game. *September 12* argues and enacts the futility of the War on Terror (you play an American bomber pilot above a Middle Eastern town, but the more enemies you kill the more they multiply); it is a game which the player can only win by refusing to play. 'This is not a game,' its instructions announce: 'This is a simulation. This is a simple model you can use to explore some aspects of the war on terror.' Frasca's *Madrid* – a sombre pause for reflection upon the 2004 train bombings in Madrid – offers one static graphic (an image of people holding candles) and has only one function and one rule: 'Click on the candles and make them shine as bright as you can.' These 'arthouse' games' modes of interactivity – their incitement to interpretation – do not rely on conventions of realism in their graphic representations, but take place at the level of a philosophical and political argument which they do not articulate so much as they allow the player to explore and perform. These games, like Barthes's *scriptible* text or Benjamin's optimal artistic experience, do not permeate and assimilate their players – instead, their players navigate, negotiate, translate and reconstruct the meanings offered by the games.

In his study of the rhetoric of digital games, Gonzalo Frasca (2007: 133) points out that in *September 12* 'the player soon realizes that the game cannot be won [...] the game sabotages itself [...] it does not provide a solution to the problem of terrorism but it aims to show that following the only allowed strategy (shooting missiles) is doomed to failure.' Frasca (2007: 17, 28) argues that while games may be seen as transmitting ideologies some (including his own) are explicit in displaying their politics. This overtness diminishes the possibilities of insidious subliminal ideological indoctrination.

Frasca (2007: 53) argues that games do not require their players' agency so much as the appearance of that agency: 'it is enough for them to believe that they are in control.' A game, says Frasca (2007: 70) is something in which the players believe they enjoy a mode of active participation. Frasca (2007: 89) notes that, while games may appear to offer agency and liberation, 'play may connote freedom but it is always constrained and, therefore, authored – either by a designer but also by the environment and cultural conventions.' Thus, Frasca (2007: 140) argues that games' design techniques guide and privilege certain performances over others, and suggests that in this way games deploy persuasive strategies to interpellate their users.

For Frasca (2007: 201) there remains a constant balance between the player's apparent autonomy and the game's regulatory structure: although the game can generate a vast array of meanings, that range is neither infinite nor extrinsic to the rules of play. One is struck by the parallels between Frasca's definition of gameplay as rhetoric and the poststructuralist philosopher Jacques Derrida's view of textual meaning as a game, one which remains within certain prescribed limits. Within this context, Roland Barthes's notion of the *scriptible* text or textual interpretation (of the textuality which invites co-authorship, or indeed of the textual approach which refutes the primacy of authorial authority and which imposes such co-authorship upon the text) seems apposite. There are limits to the extent to which we can rewrite the rules of Frasca's and Derrida's semantic gameplay, but insofar as we are critically aware of those limits – as we surely must be when we play *September 12* (in that the game draws our attention to its own problems, paradoxes and limitations) – then we can become critical and creative players or interpreters of the game/text.

But to what extent, we may ask, is this reflective co-authorship possible when the player is engaged in a commercial, mass-market digital game – a fast-moving First Person Shooter, say? A poststructuralist might propose that it is of course possible: we can actively reflect upon and thereby reconstruct even the most reactionary and recalcitrant of texts. But a pragmatist might suggest that, while this approach is clearly possible, it is also reasonably improbable. How might we establish grounds for popular negotiations with a populist game which will allow no such ground (let alone encourage such engagement) within a broader mass culture which appears unaware of the need for such agency or empowerment (insofar as the FPS game purports to have already empowered its player merely by putting a virtual gun into her hand)? Is it only in our engagement with games like *12 September* that we can reasonably hope to achieve this?

Gonzalo Frasca has argued that video games of all kinds 'require a lot of work from the player – much more work than most other media' and suggested that it might therefore be possible for mass-market games one day to offer gamers the degree of critical agency (i.e. demand the effort and commitment from the players to think critically and therefore creatively): 'It depends on the game. But I think that they could, within the right parameters. In a classroom setting, with the guidance of an educator, they can definitively do so. By themselves, it's a tougher challenge.' He has added however that he has 'no idea' whether games that encourage and require critical, creative and progressive reflection could ever be popular in a mass commercial market: 'I wish they were but there's no way for me to know this. The situation can change eventually.'

A game as challenging as Frasca's *12 September* – a game as intellectually disturbing and controversial as any aspect of *scriptible* culture should be – is Danny Ledonne's *Super Columbine Massacre RPG!* (2005), a game which, like *September 12*, offers a problematizing account of a prevalent culture of violence and its mass media representations, and of the nature and function of the digital game itself. It seems significant that all of these arthouse games overtly reject the aesthetic of graphical verisimilitude so prized by the commercial games industry: like the theatre of Bertolt Brecht, they bare their devices, distancing the player from an uncritical interpellation into the ideological worldview advanced by the game, and emphasizing the incoherent and paradoxical nature of that worldview.

There remain also, of course, possibilities that this *scriptibilité*, when not advanced by the game itself, can be seized by the player: not in the obedient gameplay videos posted on YouTube (which are no more than puppet theatre performances in which the designers pull the players' strings), but in gamers' interpretative and programmatic reconstructions of games. As Roland Barthes suggests in *S/Z* (1975) even the most conservatively *lisible* structure may be translated by radical reinterpretation into a subversively *scriptible* structuration. When Hezbollah employed Eclipse Entertainment's 3-D engine to re-imagine the likes of *Counter-Strike* as the anti-Israeli war game *Special Force*, they were following an informal tradition of constructive mis-play by which digital games players have reinvented the meanings, structures and functions of games in ways unforeseen by their designers. As Arsenault and Perron (2009: 124) point out, the processes of meaning imposed upon games by individual gamers can be profoundly different from those games' intended semantic systems. Indeed online gamers increasingly use multiplayer sites as media for communication entirely unrelated to the content of the game – as text-based social networking becomes for many users the primary function of these sites.

In 2006 al-Qaeda's Global Islamic Media Front published *Quest for Bush*, an anti-American modification of Jesse Petrilla and Bob Robinson's *Quest for Saddam* (2003), an FPS game whose absurdly propagandistic narrative and imagery almost (inadvertently) caricature its own xenophobia. Two years later the Iraqi-American artist Wafaa Bilal produced *The Night of Bush Capturing: A Virtual Jihadi* (2008) – his own deconstructive reinterpretation of al-Qaeda's modification of Petrilla and Robinson's original game. There is both political and artistic scope within this hacking or hijacking of commercial digital games to sponsor new processes of meaning; and yet these radical misreadings (by producers or by users of games) remain relatively rare, just as the Postmodernist art game (despite its academic champions) continues to languish in commercial obscurity. While the potential remains for any gamer to reconstruct the meanings of a game, the ideological, economic and aesthetic principles of commercial games production tend to limit these heteroglossic possibilities.

It may not be the inherent nature of the digital game itself so much as the historical context of its development and popularization (the axes of late postmodernity, globalization and the War on Terror) which has prompted contemporary cultural philosophy to view it as the epitome of the superficial and the homogeneous, of the post-material hegemony of the simulacrum. There is, after all, nothing more virtual or immersive about the reality of *EverQuest* than that of *David Copperfield* or *Citizen Kane*. Nor is there anything extraordinary or unprecedented in the ways in which the economics of the digital games industry have inhibited the medium's creative and political potential, and simultaneously exploited its revolutionary or liberatory reputation for interactivity.

This illusion of interactivity disguises the video game's actual USP: the potential for the gamer to assume a manufactured subjectivity – not for the self to interact with the game, but for that self to be subsumed to the game's constructed subject. The commercial video games industry has put an extraordinary emphasis upon the significance of interactivity in its field of production; but while the digital game may be neither more nor less interactive than any other text, this emphasis itself seems not only to mislead its consumers but also to disempower them. When, then, the cyber-optimists' dreams of a world of *World of Warcraft* are eventually realized, this new reality may not represent the democratic cybertopia of which some have dreamt. W.H. Auden (1979: 81) suggested that 'each in the cell of himself is almost convinced of his freedom' – yet when each individual is not *almost* but *absolutely* convinced by that illusion of self-determination, those who are interpellated by the dictatorship of the algorithm will not even dream of autonomy and liberation, because (like all those sustained by the *faux-scriptible*, like the victims of *The Matrix* itself) they will mistakenly believe that they are already the authors of their own destinies.

Transfigurations

In their account of social interaction within virtual environments, Cheng et al. (2002: 98–99) note that, while most participants in their study used their avatar to convey facets of their real-world identities, others deployed their avatars as expressions of wish-fulfilment, representing themselves with images of famous or attractive figures. In their study of the uses and meanings of avatars in social media, Vasaloua et al. (2008) also noted that, in choosing or constructing an avatar, users may select options most closely aligned to own perceptions of their own attributes, while often at the same time idealizing their avatars by either concealing or emphasizing particular aspects of themselves. The avatar then comes to be a projection of desire, a reinvention of self in accordance with a mode of desire which we see as emanating from within each subject but which may in fact be imposed upon the individual subject by its socio-cultural conditions of existence, by what the sociologist Pierre Bourdieu might describe as its *habitus*.

James Cameron's 2009 film *Avatar*, like the video game *America's Army*, offers a vision of the War on Terror fought through the tools of virtual reality – one in which the avatars designed to capture the 'hearts and minds of the natives' eventually overwhelm their users. The film's hero discovers himself becoming his avatar: 'everything is backwards now – like out there is the true world and in here is the dream. I can barely remember my old life. I don't know who I am anymore.' Thus Cameron's hero chooses to surrender his old life and his old body in favour of the virtual and fantastical world in which he has found himself – or, rather, in which he has reinvented himself – or, rather, in which his self has been reinvented.

Cameron's hero's subjectivity becomes at once liberated by and subsumed to the existential condition of his avatar. What Dixon (2007: 241) has called the 'digital double' is always, eventually, an uncanny double, a doppelgänger which, as Freud (1985: 335–376) suggested, returns to question the integrity of the original self. The projection of the self comes to expose the subject's lack of autonomy, insofar as the projection (a product of mass mediation) at once overwhelms the original subject and reveals the fact that this subject has never itself been anything more than the product of such mass mediation.

The *Oxford English Dictionary* defines an avatar as 'the descent of a deity to the earth in an incarnate form' – as a 'manifestation or presentation to the world as a ruling power or object of worship.' Yet, while Cameron's futuristic film might finally allow its protagonist the physical possibility of that apotheosis, the virtual avatar of contemporary media affords its user only the illusion of that divine power – whether as a player of digital games or as the viewer immersed in James Cameron's 3-D world. This is not the incarnation of a god in human form but the illusory manifestation of a human in the image of a god or ruling power. The pseudo-utopia of the cinema, the video game or the Internet does not as such represent a process of empowerment but one of consolation.

The *Oxford English Dictionary*'s definition of the word 'agent' is also perhaps worth considering in this context: 'one who [...] acts or exerts power [...] he who operates in a particular direction [...] one who acts for another, a deputy [...] substitute, representative or emissary [...] the material [...] instrumentality whereby effects are produced; but implying a rational employer or contriver.' When the subject considers themselves an agent then they may identify themselves as one displaying agency or autonomy; and yet an agent is also a functionary or operative of an agency or authority. The hero of James Cameron's *Avatar* transcends his status as an operative of military-industrial authority to achieve some level of autonomy (he will not serve) – but this is of course a matter of fantasy, a wish-fulfilment which interpellates subjectivity and sustains ideological subjection. If we are all *secret agents*, then the defining secret (the secret in which we are complicit in hiding from ourselves) is that we are agents in the subservient rather than the autonomous sense of the word. That is not to suggest that there is a rational contriver (any more than it is to require an intelligent designer in order to explain evolution). As Foucault (1991) and Bourdieu (1977, 1986) have supposed, such self-perpetuating, self-evolving structurations seem more a case of unconscious complicity and consensus than of conscious conspiracy.

Bourdieu (1977: 85) depicts the process of mass culture as a work of inculcation by which history's objective structures perpetuate themselves through human individuals. This process, for Bourdieu (1977: 79), involves a mediated illusion which allows the individual human agent's actions to appear to be the product of reason. This consoling notion of the ministra-tion of conscious and reasonable agency (of the individual subject, of an autocratic elite, or of a divine architect) – what Bourdieu (1991, 2005) has called the *mystery of ministry* – is only as comforting as it is successfully illusory. Despite all the evidence, rational agency remains an idea to which we relentlessly cling.

An experiment conducted by Soon et al. (2008: 543) has demon-strated that 'the outcome of a decision can be encoded in brain activity up to 10 seconds before it enters awareness.' These neuroscientists hooked up a test subject to a brain scanner and asked her/him to press one of two buttons: the subject was asked to press the button as soon as s/he had decided which button to press. As early as ten seconds before the subject pressed the button, the scientists were able accurately to predict (based on which areas of the brain were shown by the scanner to be active) which button the subject would press. The implications of this experiment are somewhat disturbing: that the brain makes decisions before the conscious mind becomes aware of those decisions, and that therefore one's conscious-ness, one's subjectivity, is not in control: subjective agency (or 'free will') is an illusion.

We might go further and envisage an extension of this experiment: imagine that we tell the test subject that they should themselves choose which one of the two buttons to press unless (before pressing the button) they receive an instruction from the scientist as to which button to press (they should then press the button as instructed); the scientist then tells them to press the button their brain has already decided upon (although this decision has not yet filtered through to their consciousness); they therefore press the button in accordance with their own brain's decision, but in the belief that they are doing so in response to the scientist's external agency. This exposure of the deterministic nature of being advances an existential alienation whose disillusionment may be the closest thing to empowerment that we can hope to achieve – insofar as our only empowerment can be the recognition of the irrevocability of our disempowerment.

Yet what seems to us to be most empowering is, for the most part, the very antithesis of this: not the recognition of our disempowerment but the comforting illusion of our empowerment. This illusion of effective agency is something which digital games, or for that matter reality TV shows, offer us in abundance.

CHAPTER 4

Reality Television

Reality television offers interactivity to its audiences: the opportunity to immerse themselves in the 'real' social interactions of 'real' people, and very often also the experience of direct interaction through voting for their favourite or least favourite contestants. In its most popular form (in such series as *Big Brother* and *The X Factor*) reality television has become an ostentatiously interactive multimedia experience, courting audience participation online, by telephone and through the use of the interactive functions of digital broadcasting. This chapter will however contend that this sense of interactivity may be as illusory as the 'reality' which these programmes purport to reflect.

One of the most insidious and exploitative aspects of purportedly interactive mass culture is, as Papacharissi (2010: 65) suggests, the way in which 'the guise of participatory media and the promise of power' entice audiences to produce content without ever being compensated for their work. Yet there is an illusion of compensation: the magnanimity of the process by which the medium offers to elevate and empower its participants, the gift of symbolic capital which it confers upon its subjects and (both vicariously and by offering a simulacrum of democratic or creative agency) upon its audiences. But, as Pierre Bourdieu (1977: 195) has argued, this gift of symbolic agency is one which, insofar as it claims to be both free and liberating, silently binds its subject within its established power relations: 'a gift which is not matched by a counter-gift creates a lasting bond, restricting the debtor's freedom and forcing him to adopt a peaceful, co-operative, prudent attitude.' This process of ideological assimilation and hierarchical entrenchment necessarily disguises its repression beneath a veil of liberation. Repression masked as liberation through an illusion of agency affords and manufactures the consensus essential for the reinforcement

and perpetuation of power structures – inasmuch as, again in the words of
Pierre Bourdieu (1977: 195), 'the endless reconversion of economic capital
into symbolic capital [...] which is the condition for the permanence of
domination, cannot succeed without the complicity of the whole group.'

This phenomenon is seen clearly in the proliferation of reality television; through, for example, the divine generosity of Big Brother or Simon
Cowell. Tincknell and Raghuram (2004: 263) have, for example, argued
that 'the *idea* of agency' was crucial to the success of the ground-breaking
series *Big Brother* – but that this should not imply that the audience was
granted any real agency or textual autonomy:

> the extent to which this constituted the power to determine the meaning produced
> *in* and *by* the text is debatable. The production company [...] certainly emphasised
> the importance of audience participation [but] clearly retained editorial control of
> what was seen and heard on the programme.

In 2008 the media critic and journalist Charlie Brooker authored a
television horror series called *Dead Set*, a drama which depicted a world
overtaken by zombies – witnessed from the perspective of the residents of
the *Big Brother* house. Brooker's series portrayed a contemporary society
of mindless ghouls drawn to the *Big Brother* house, a devastated society
whose short-lived human survivors enumerate the landmark achievements
of their culture as celebrity magazines, junk food, popular television and
social networking and shopping websites: *Heat, Nuts, Zoo*, McDonald's,
Nando's, *EastEnders*, YouTube, Facebook, MySpace and Amazon. As the
Big Brother contestants survey the post-apocalyptic landscape, one asks:
'Does this mean we're not on telly anymore?' When television has become
the yardstick of reality, the end of the world is only meaningful insofar as it
also signifies the end of television. To paraphrase a famous adage of Slavoj
Žižek's, we might suppose that it is in fact easier to imagine the end of the
world than to imagine the demise of mass-mediated capital.

As Geoff King (2005: 94) proposes, the paradox of television – as of
the video game – is that it has come to seem at its most realistic when it is
in fact at its most mediated. Jean Baudrillard (2005: 75) suggests that in the
simulacrum of reality television – as in the video game – we are witnessing

'the confusion of existence and its double' – a confusion between reality and representation, and a prioritization of the representation of material existence over its reality – a confusion then, between reality and television.

The banalities of reality television have thus come to seem more significant than events we might once have considered of political or historical weight. In July 2010 Australia chose to reschedule a televised election debate from its traditional slot in order to avoid a clash with the reality TV cooking show *MasterChef*. The political has clearly here been subsumed to the demands of the mass entertainment industry. One might note in passing that Australia's version of *MasterChef* has developed sufficient public significance that in July 2011 the programme managed to attract the Dalai Lama to act as a guest judge. On 19 July 2011 the *Daily Mirror* reported that 'stunned contestants prepared lunch for the Tibetan spiritual leader – but he refused to rate their offerings, saying it would be against his Buddhist principles.' Meanwhile the British TV chef Jamie Oliver has become known for his political campaigning (specifically his campaign for the provision of better food in schools); and in spring 2012 it was revealed that fans of Egypt's popular TV chef Ghalia Mahmoud had asked her to stand for the nation's presidency.

In May 2009 Piers Morgan, a former editor of the *Daily Mirror* and a judge on the popular talent contest *Britain's Got Talent*, described one of that programme's contestants (a Scotswoman whose unprepossessing appearance had garnered her international fame) as an antidote to global recession. In doing so, Morgan invoked a typically mediacentric perspective upon contemporary society, one which promotes mass-mediated image above material substance and which (in the era of the spin doctor and of reality television) seems virtually unassailable. As Biressi and Nunn (2005: 144) have reminded us, the proliferation of reality television has coincided with a period in which politics has been increasingly packaged as media product – and it seems that just as politics has blurred into the field of popular entertainment, so entertainment has assumed a new socio-political significance, or has transformed itself into a substitute for that significance. Indeed, Ouellette (2009: 240) has gone so far as to suggest that reality television may be seen, in some of its manifestations, as an active agent in the neoliberal transformation of democracy.

House of Cards

Maitles and Gilchrist (2005) have pointed out that more people voted for the winner of Channel 4's *Big Brother* than the combined vote for the Scottish Parliament, Welsh Assembly and London Mayoral elections between 1999 and 2000. Kilborn (2003: 15) meanwhile asks what it tells us about our culture that more people vote on *Big Brother* than in European elections. This late postmodern, post-postmodern or post-historical era appears to be witnessing a process whereby material history is increasingly subsumed to a mediated virtuality, to the reality of television, of reality television, a transformation of the material and the substantive into a mass media product. This much has been painfully apparent since (in January 2006) British MP George Galloway (who had in 1994 and 2002 launched his own peace missions to Baghdad) appeared on *Celebrity Big Brother* in a leotard and pretended to be a cat – or since (in January 2007) racial tensions on the same show prompted the burning of effigies of one of its contestants in the Indian city of Patna and apologies from the then Chancellor during a visit to India – or since (in July 2007) the then Prime Minister was obliged to comment on that franchise's next race row – or since (in March 2009) the next Prime Minister the tributes to the late *Big Brother* star (and former alleged racist) Jade Goody.

Stephen Coleman (2006: 457) has helpfully described *Big Brother* as representing 'a counterfactual democratic process in which conspicuous absences in contemporary political culture are played out.' In response to the reduction in political trust provoked by a perceived democratic deficit, Coleman (2006: 477) indeed goes so far as to suggest that reality television producers might imagine media formats that restore public interest and participation in politics. This perspective appears to be shared by politicians (including Barack Obama and David Cameron) who have courted such reality television gurus as Simon Cowell.

There appears, as Putnam (2000: 235) suggested, a close correlation between a society's dependence upon television and its tendency towards civic disengagement. It seems unclear, however, whether a public decline in political engagement merely happens to coincide with the rise of reality television, or may in fact be related to it – insofar as these TV shows' simulacra of participation have come to sublimate increasingly frustrated popular desires for democratic agency. Furthermore, one of the problems repeatedly encountered by the idea that developments in media technologies and formats can be effectively harnessed for the promotion of democratic participation is that these tools are for the most part employed by governments, political parties, political and commercial interest groups and individual politicians not for the sake of the reinvigoration of the public sphere but with their focus fixed resolutely upon the reinforcement of their own specific interests and bases of power. Thus, as Blumler and Gurevitch (1995: 221) have warned, attempts to manage public communication in order to manipulate the public have resulted in the alienation of citizens from democratic processes – the very trend of political exclusion which such strategies might ostensibly be intended to reverse.

Kaur (2007: 11) cites former *Big Brother* contestant (and former Conservative parliamentary candidate) Derek Laud's reasons for appearing on that show: that, while increasing numbers of people were registering their opinions in reality television votes, decreasing numbers were voting in general elections. According to Jonathan Bignell (2005: 96) the British Member of Parliament Jane Griffiths contacted programme-makers Endemol in November 2003 to suggest the launch of a House of Commons version of *Big Brother*. It seems that Ms Griffiths believed that in the hearts and minds of the great British public the political significance, responsibility and privilege of parliamentary democracy had already been subsumed to the pseudo-democratic simulacra of reality television. Jane Griffiths has since added that she 'thought it would make MPs seem more human.' It appears that we need the fantasies of reality TV to make political reality look real.

Channel 4's 2010 series *Tower Block of Commons* – in which members of parliament shared the lives of the residents of British council estates – to some extent explored this format, but, insofar as it promised to focus more upon the experiences of the tower block tenants than upon the personalities of the politicians (who interacted with the tenants but not with each other) it offered to anchor its discourses closer to socio-economic reality than to the depthless representational matrices of reality television. One participant in the series, Liberal Democrat MP Mark Oaten has commented that 'getting politicians away from Westminster, the *Today* programme and party political broadcasts is one way of reaching out to more people. Living in a tower block helped me contact people who never listen to conventional political debate and it taught me a thing or two.'

Another participant, Conservative MP Tim Loughton added: 'If done properly like *Tower Block of Commons* then I think these programmes have the potential to engage more people in the political process who otherwise would not be interested. It also has the potential to challenge common misconceptions about all MPs being out of touch. I received hundreds of emails from complete strangers from around the country saying how the programme had completely changed their view of MPs and Conservative ones in particular. I doubt you would get the same response from Austin Mitchell though!' Indeed, Labour MP Austin Mitchell, another of the politicians who appeared on the programme wrote on his blog in February 2010 that he regretted his participation in the series, which, he claimed, set out to 'humiliate MPs' – adding that the series was a 'disgrace' and describing those responsible for it as 'bastards'. Mitchell argued that the series fitted the superficial stereotype of reality television by concentrating more upon the characters of the politicians involved than upon the conditions experienced by residents of public housing estates.

Austin Mitchell was also kind enough to comment on his participation in *Tower Block of Commons* for the purposes of this chapter. As not only a reflection on the state of reality television, but also a depiction of the experience of disempowerment involved in participation in the genre, his comments are worth including at length:

> Reality TV is currently fashionable for obvious reasons. It's cheaper than studio-based programmes. It eliminates tired, traditional formulae and overpaid presenters. It's real in the sense that it involves real people (i.e. exhibitionist ones). The current fashion for reality TV has inundated a population living in a deteriorating reality with escapist programmes.

Reality programmes on social issues can be a major force for raising concerns and exposing a comfortable audience to issues of concern and realities of which they'd otherwise know nothing. Ken Loach's 1966 TV drama *Cathy Come Home* was not reality programming but it showed the enormous potential for generating concern and producing action by driving home the realities of homelessness. Reality programmes have the potential for doing the same for poverty, alienation, dependence, low wages and exploitation, all brought home through honest portrayals of the lives of the people exposed to them. The one requirement is that the portrayal be honest and, therefore, trustworthy. The basic issue is integrity.

Which is the reason why I was upset and annoyed at *Tower Block of Commons*. I thought I was participating in a programme which would show the reality of what it's like living in tower blocks. The producers were actually intending to make fools of MPs by showing that they can't handle the realities of hard lives, which I freely admit I can't. As I see it, they wanted to show MPs as out of touch, high-living figures of fun. Looking back I was daft not to think that this would be the intention, and lots of people certainly warned me about this. Probably because I was too keen to take part.

The producers assured me when they took me out to dinner when planning *Tower Block* that this programme would be a serious and concerned look at the lives of the people in the blocks. The MPs would be on a voyage of discovery to see this and project their concerns to a wider audience.

Here was something I was keen on, being Chair of the Commons Council House Group, and an active campaigner for a fairer deal for council houses, because I was very concerned that tenants of tower blocks are being forced to live in intolerable conditions because of the government's failure to spend the money necessary to refurbish and bring the blocks up to Decent Homes Standards and the resulting use of them as social dumping grounds for poverty cases and people with problems. Britain has a major housing shortage and lots of tower blocks so it's important to improve them rather than pull them down, and it seemed to me that if the realities of life there could be honestly portrayed on television then it would drive home our case for improving their lot. The producers assured me that this was their intention. This was to be a programme with a mission.

Then the problems began. The director wasn't allowed a mind of her own and was insistent on following the brief and daily instructions given her by her bosses in London. 'I've got a mortgage to pay,' she kept telling us when I refused to jump through the hoops she'd been sent up to put me through. A nice lass but one on tight remote control, she immediately set out conditions which hadn't been mentioned before. These were that I should wear what she described as 'estate uniform' namely T-shirt, tracksuit bottoms and plimsolls. I didn't see any other pensioners in Orchard Park in this kind of outfit and I looked ridiculous in it so I refused. Which upset her and meant lots of questions thereafter from the people they introduced me to on the estate, like why wasn't I dressed like them? Since the other people in the programme could hardly have been bothered about this, I think they were put up to it.

The other requirement was that I should live on £100 a week which I'd have got on Job Seekers Allowance. I didn't see this as a programme showing me struggling to live on JSA when the focus should have been on how real people managed. I refused this too. Which meant that everyone I was introduced to devoted a lot of time and questions asking me why I wouldn't live on £100 a week. Again I feel they were put up to this.

They'd found us a flat by paying the girl living there to stay with friends for two weeks, but it had no bed, the oven wasn't working, there was a welcoming turd in the toilet, and the place was a mess. We collected some furniture from the Salvation Army and assembled the bed but most of the furniture wasn't delivered, we couldn't cook and there was no crockery, so we phoned friends in Hull and borrowed their garden furniture. They invited us round that for a meal for that first night. We went but the producer insisted on filming this sensible move as if it were as an escape from the housing estate. It wasn't, merely hunger, exhaustion and a chance to find out about the housing situation in Hull. We went back after the meal to the flat with the producer trying to persuade us to stay in her hotel (though warning that she'd film it). I wasn't going to cop out so we refused, put the bed together after a struggle and went to sleep. But not to dream, for that first night revealed how serious the problems were. None of this was filmed but the door bell rang several times in the night as people tried to get in by ringing everyone's bell in the hope that someone would open the door. This happened every subsequent night making sleep difficult, particularly when one resident started riding his motorbike round the internal corridor and then up and down the lift.

The programme was a series of missed opportunities. I asked regularly why we saw nothing of the other residents in the tower block, why we didn't explore the shops (few and fortified) or the schools, or the lack of facilities for young people. No time for any of that, apparently, but plenty to waste showing me shopping with no idea of prices or me cooking incompetently, changing nappies, and learning how to place bets (none of which are part of my parliamentary duties and none of which I ever do), all totally irrelevant and totally useless so far as the main purpose of the programme, the life of the people, is concerned.

Who cares about me?, I kept asking, but the director had a brief to follow. It was dictated from London and she wasn't allowed to vary it for local needs or issues.

The producers were more intent on putting MPs through humiliations than focussing on the people who have to live in tower blocks. Had it been presented at the start as a programme about MPs and how they could adjust to council house conditions and benefit levels then I might not have taken part because I wanted to show how people live in tower blocks, not make a fool of myself. But if I had then taken part (though handicapped by incompetence and age) I'd have approached it in a totally different fashion and given the tenants a far better say. In the end they didn't even get paid for allowing their time to be wasted. I find this appalling. I wasn't paid but then I don't need the money. The people I met in Orchard Park really do.

This doesn't of course degrade the whole category of reality programmes. Honestly done, they can (and should) reveal the conditions and views of a whole range of social groups, let them speak for themselves and put their case to a wider audience. They could bring diverse groups and voices onto the public agenda. That's a democratic advance. It's not real power but it's as close as a lot of people will get.

The speeches of politicians are words spoken into the wind: ephemeral and unheard unless they are illustrated by real problems and messages from real people. Reality TV can do this. I see it as a force for social concern and improvement because that's my aim, but it's also useful if it educates by showing us the views and conditions of other sections of society. We're all too segmented and self sufficient these days but reality TV can bridge the gaps, spread information, illustrate other lives and give a voice to the silent sufferers. That's educative and informative.

However, now that we are in a multi-channel situation, the audience channel-hop around far more than they used to in those days of channel loyalty. This means that producers have to introduce an element of entertainment, or even shock, to get the audience.

Mitchell makes a number of key points: that reality television could clearly function as a powerful instrument in the progressive development of society (it is obviously influential upon society), but that in order to do so it would need to reflect reality rather than – as Fiske (1987) and Herman and Chomsky (2002) might suppose – manufacture or fabricate the real world. In order to do so, it would need to shift its focus from an overweening concern with image (the superficial, the virtual) to a concentration on real material issues. Yet this focus upon image is clearly a phenomenon which influences much contemporary news-making (see, for example, McQueen 1998), and also central to the practices of modern politics (see also, for example, McLuhan 2001, Greenstein 1967, van Ham 2001 and Savigny and Temple 2010). As Swanson and Mancini (1996: 272) have suggested, the dominant contemporary political strategy involves the personalization of telegenic leaders and the sidelining of policies. Or as Jean Baudrillard (2005: 98) has put it, 'the image is more important than what it speaks of.' Within this increasingly virtual culture, the prospects might therefore seem bleak for Austin Mitchell's hope that a return to social and material realism might restore mass media entertainment's progressive, campaigning role. Reality television may rarely reflect material reality, but, like other aspects of contemporary virtual culture, it may transform that reality into its own depthless veneer.

The X in the Box

In his study of reality television Jonathan Bignell (2005: 130) details a particularly disturbing instance of the confusion between reality television and historical reality – when, on 11 September 2001, the producers of the American version of *Big Brother* called one of the programme's housemates into the diary room to inform her (in the anonymous, invisible voice of 'Big Brother') that her cousin who worked in the World Trade Center was among the missing. This confluence of reality television with the major events of material history reminds us that reality television has itself become one of those major events. In an article in *The Sunday Times* on 3 January 2010 Rod Liddle pointed out that in its review of the first decade of the twenty-first-century broadcast a few days earlier, on New Year's Eve 2009, BBC television 'cut from footage of those planes smashing into the Twin Towers to a man called Will Young winning the original incarnation of *The X Factor* – as if these two crimes against humanity were equal in their devastation.' It seems at least that these events are increasingly accorded similar historical significance by the mass media.

In May 2006 *The Observer*'s television critic Andrew Anthony asked:

> Where were you when Shahbaz walked out of the *Big Brother* House? It's not quite the Kennedy assassination, I'll concede, but we can't choose the gravity of the times in which we live.

Reality television has come to replace material history; events are only and essentially media events. In August 2006 Andrew Anthony added that *Big Brother* 'fulfils its Orwellian promise: it *is* society.' Multimedia television is the new society, a post-historical substitute for politics and democracy. When, for example, in January 2010 Tony Blair faced the public enquiry on the Iraq War, *The Daily Telegraph* announced that (in the style of a reality TV show) its online readers would be able to vote live on the veracity of every statement made by the former Prime Minister – with a truth meter on the side of the screen displaying (like a TV talent show's vote counter) the audience reaction to Blair's testimony: 'You'll be able to follow a live stream of the testimony as it happens and give us your opinion by voting on our live lie detector, which follows votes to track whether the viewers believe him or not on any particular issue.' Indeed coverage of the UK's televised election debates a few months later included live audience response meters displaying the ups and downs of the three leaders' popularity ratings as the debates progressed.

Reality television, like contemporary politics, has transformed issues of substance into matters of image and performance. Reality television does not, as its name might suggest, unintrusively observe ordinary people in ordinary situations: that is the province of fly-on-the-wall documentary. On the contrary it takes extraordinary people (extroverts and exhibitionists, the remarkably talented and the remarkably talentless, eccentrics and celebrities) and puts them into situations to which they are unaccustomed and which to them are extraordinary. The reality of reality television is therefore radically artificial; it derives its illusion of reality from an aura of immediacy founded at first upon a vicariousness or intimacy of association and later upon a promise of audience participation or interactivity. The popular British television series *The X Factor* (2004–), for example, closely fits this model. It offers members of the public the opportunity at once to achieve fame and fortune (to win the celebrity lottery), to share the vicarious pleasure of the achievement of that fame and fortune (through identification with the contestants) and to influence the contestants' progress (through a telephone voting system – which also offers its users the chance to win tickets to join the studio audience, to participate even more closely in the show – as well as to win a life-changing cash prize). It is, as such, a traditional talent contest which also affords the modes of pleasure associated with such archetypes of reality television as *Big Brother*: liveness of performance and immediacy of reaction, vicarious achievement and a

sense of democratic participation. Yet even this illusion of participation –
this mock-democracy, this democracy of insignificance and ephemera – is
limited. Until the show's final stages the weekly public vote determines
two candidates for expulsion from the process: the programme's judges
then decide which constestant to reprieve and which to discard. (It is as
if I repeatedly asked you to choose two cards to reject from a pack, and
then just rejected one of these; you would enjoy the illusion of agency,
but we could get through the whole pack and I would still have left in my
hand the one card I had decided to keep from the start.) Winning is in
many ways less important than merely reaching the final rounds of such
series: such finalists who failed to win as Susan Boyle, Olly Murs, Jedward,
and Gareth Gates have often gone on to highly successful careers, often
managed by the show's judges, while some actual winners (such as Steve
Brookstein and Leon Jackson) have slipped into relative obscurity. Indeed
those who appear to melt the heart of the overtly cynical Simon Cowell
tend to stand the best chance of stardom under the great man's patronage
(and what could be more cynical than that?).

The X Factor's celebrity judges in this way continue for the most part
to have the final (or the most important) say: democracy only extends so
far; the ultimate power continues to rest in the hands of a self-appointed
elite, an oligarchy. There is perhaps a comforting conservatism in this situation which appears to appeal to audiences excluded or alienated from more
traditional structures of social and political trust.

A contestant in the 2011 series of The X Factor (who preferred to remain
anonymous) commented that this sense of participation seemed illusory.
It was not only that, as one might imagine, the frenzy of the live studio
audience was clearly manufactured: 'they tell the audience to scream, they
tell them to clap, it's all on cue.' She also felt that even the judges' own ver-
dicts on contestants had been pre-determined: she commented that, when
she received her four 'yes's from the panel, it seemed staged or agreed in
advance: 'A lot of it has already been set up. It's set up for the reaction that
they want from the judges. It's like they got told to build the tension around
us.' One might suggest, then, that the judges do not participate as sponta-
neous individuals so much as they function as agents of the programme's
dramatic structure. Like the President, Queen, General, Archbishop or Vice

Chancellor (like all those whose names and individual identities are subsumed to their titles, their positions) – like judges in the courts of law (and this is why such judges have traditionally worn the regalia of their offices: in order to deny their individual identities) – these talent show judges perform institutional roles which exist and are empowered before and beyond their status as individuals (a status also subsumed to their personal images as manufactured by the gossip magazines and the red-top press). This is not to undermine their power; on the contrary, it incontrovertibly reinforces the power for which they act as vessels. They are no longer mere mortals: they have assumed a mantle of institutional power, or, rather, that institutional power has assumed the mantle or guise of these individuals. Thus, kings and queens, and judges and ministers, may come and go (even its founding father Simon Cowell may take a leave of absence from *The X Factor*), but the power structures remain. Pierre Bourdieu (1977) sees such institutional roles as necessarily distinct from the individuals holding them – which allows for them to be occupied by different but equivalent individuals. Thus the power structure is no longer reliant on the endurance of the individual; as Bourdieu (1977: 187) suggests, 'once this state of affairs is established, relations of power and domination no longer exist directly between individuals; they are set up in pure objectivity between institutions.' This allows for an agglomeration of power beyond the reach or the span of the individual.

In an episode of November 2009 *The X Factor*'s presenter Dermot O'Leary announced that, as the series had reached its quarter-final stage, the judges would no longer have the power to save acts from expulsion: 'Judges, you are powerless. This is how normal people feel all the time.' Yet, of course, the judges retain their predetermined positions of influence: their judgments continue to inform the perspectives of the public jury. O'Leary's differentiation of the programme's judges from 'normal people' far from undermining their authority in fact underlines it. They are not 'normal people' insofar as they are not individuals at all: they are performing institutional roles. But they are 'normal people' in that they are powerless as individuals within these structured roles. The matters of personal taste which appear to motivate their choices are merely, as Pierre Bourdieu (1986) would argue, issues of socio-economic distinction. The tastes of the individual (judge and viewer) are determined in advance by their cultural identification, by what Bourdieu would call their 'class'.

The results of the early weeks of audience votes in the 2009 series appear to demonstrate a number of interesting trends. During the first four weeks of these public telephone votes, four of the six acts whom the viewing public afforded the fewest votes included performers of ethnic minority (one of whom appeared three times in this least popular category in the space of a month); while only one of the six acts not to face such public disapprobation featured a non-white performer. During the first six weeks of public voting four of the six acts evicted from the show featured ethnic minority performers – while only one of the remaining six acts featured a non-white singer (who was eventually expelled in the semi-final). A similar effect was witnessed in the BBC's 2008 talent competition *I'd Do Anything*, in which the only two finalists of ethnic minority (out of a cohort of twelve) – Cleo Royer and Keisha Amponsa-Banson – were voted off in the second and fifth weeks of the series respectively. The latter faced the final sing-off three times, having received the lowest number of public votes on each occasion. This situation eventually (on 27 April 2008) prompted an unprecedented outburst from chief judge Andrew Lloyd Webber: 'this is a complete and utter travesty – for the first time on a television show, I am angry.' (See also Sherwin 2012.)

These voting patterns may not suggest an overt or conscious racism prevalent among the programme's audience; it may simply be that the voting members of the audience (like those in the *Eurovision Song Contest*) tend to vote along lines of class, ethnic or cultural identification – a position more closely aligned to jingoism than to xenophobia. This process of identification would appear to reflect the vicariousness of the pleasures the programme affords. However, the successes of Leona Lewis and Alexandra Burke in previous seasons of *The X Factor* suggest that this bias does not necessarily dominate the later stages of the series, when familiarity, quality and months of mediation may override more atavistic modes of identification.

On 25 October 2009 *The X Factor*'s presenter, as usual, announced the acts selected by the popular vote to return to sing the following week. As he did so, the camera turned to the programme's contestants, cutting from anxious face to anxious face, and a pattern slowly became apparent:

in four out of eight cases (indeed in alternating cases) the name announced as reprieved from expulsion would be that of the second contestant to whom the camera had just cut in each of these short montages. This half-pattern could not, of course, become too obvious (if it happened every time, rather than alternating times, the audience would notice); but it no doubt afforded the viewer an unconscious hint of whose name was about to be announced. The *X Factor* results shows of 18 October and 1 November 2009 offered similar patterns: in (respectively) six out of nine and six out of seven cases, the contestant to whom the camera had cut second in the sequence of anxious faces immediately prior to each announcement turned out to be the next one reprieved by the public vote. Watching this, the viewer (before becoming conscious of the pattern) may have discovered that she was successfully predicting the results at least fifty per cent of the time. This of course adds to the viewer's pleasure, because it enhances her illusion of agency.

Elizabeth Day (2010: 22) has written that 'part of the attraction is the sense of control the *X Factor* gives us: the sense that we can put right wider social wrongs by voting for our favourite contestants' – and that this illusion allows the programme's audience 'a much-needed sense of agency.' Day (2010: 20, 22) has further suggested that *The X Factor* affords a sense of community engagement lost to the denizens of 'a world increasingly dominated by Facebook and Twitter' – while noting therein the irony that reality television is itself a similarly virtual construct.

The success and the pleasure of the classic reality television format appears to rely on three key elements: the semblance of reality (the liveness of performance and response, the openness, normalcy and naturalness of performers' personalities, identification with those performers or contestants and the sharing of their heartbreaks and triumphs), the feeling of active participation (through telephone voting to save those contestants you like or to evict those you loathe) and a sense of the significance of the event. The last of these requires the collaboration of other media forms and outlets: *Big Brother* and *The X Factor* become significant events only insofar as tabloid newspapers, celebrity magazines, 'soft' news programmes and entertainment websites recognize and represent them as such – and also insofar as they feature increasingly within the context of more serious journalistic content. If the first rule of reality television is that you must talk about reality television, then the second rule of reality television is that you *must* talk about reality television. Even the nation's Prime Minister has colluded in this process. In November 2008 *The Daily Telegraph* revealed that Gordon Brown was a closet fan of *The X Factor* after it emerged that he had sent personal letters to the show's contestants. In November 2009 Brown, while still serving as Prime Minister, informed *GQ* magazine that he was an *X Factor* fan – and a few days later told Manchester radio station Key 103 that he was not fond of one particular act on the programme, the teenaged Irish twins John and Edward Grimes (aka 'Jedward'). Brown's opinions were discussed by the programme's judges on the 7 November edition of the series, in which showrunner Simon Cowell asked fellow panellist Louis Walsh to apologize to the Prime Minister for implying that his views on the Grimes twins suggested he was out of touch with the mood of the nation. However, on 22 November 2009 the BBC reported that Gordon Brown had sent his best wishes to the twins in response to negative reactions he had received to his criticism of their performances.

Earlier that month *The Daily Telegraph* had reported that John and Edward Grimes has 'a new celebrity fan' – David Cameron. The *Telegraph*'s use of the term *celebrity* to describe the Conservative Party Leader and future Prime Minister is revealing in itself. This confusion of political reality with television celebrity was further emphasized when shortly afterwards the Labour Party's PR machine issued an online poster which showed David Cameron and his then shadow Chancellor George Osborne morphed into the Grimes twins – alongside the slogan 'you won't be laughing if they win.' It appeared that loyalty to this series have become a significant test of one's political mettle. Although it is unclear to what if extent (if any) these events impacted upon long-term voter choice, it is perhaps worth noting that the week after the height of this controversy an ICM opinion poll showed that the trend of Conservative gain had been briefly reversed when Labour narrowed the gap between the political parties by four percentage points.

Later that same month, when Baroness Ashton was appointed to the position of the European Union's first ever High Representative for Foreign Affairs and Security, she informed the British press that she was an *X Factor* fan. It seems that this series also represents an essential point of cultural reference for activists far beyond the political mainstream. In October 2008, the radical Islamic cleric Omar Bakri provoked a minor terror alert when he announced that *The X Factor* was anti-Muslim for releasing a charity single in aid of wounded British troops.

Each episode of *The X Factor* opens with a glossy title sequence in which a shining 'X' shoots across the solar system towards the Earth, zooming down upon the UK like a falling star. One is reminded in this context of the Christian nativity – or, for that matter, of the opening of Leni Riefenstahl's *Triumph of the Will* (1935) in which the Führer descends from the heavens like a Messiah. These opening titles remind us that this is a pivotal moment in the history of the world: the genesis of an avatar, a divine star.

The reality television show thus announces to its audience that it offers an event which is extraordinary and fantastical and at once both real (historical) and significant (historic): one in which the audience can participate both vicariously and actually. It offers its audience a remarkably effortless way to assert their subjective significance by participating actively in the historical process – to have the illusion of making history – inasmuch as this is precisely what history now is, insofar as this is how power (media power, which is public power) now views history. The most successful reality television shows offer themselves as historically significant by virtue of their ubiquity, both as global media formats, and as national media events (and the complicity of the popular press is clearly key in this): Hill (2005: 4), for example, notes that half Sweden's population watched the finale of the Swedish version of *Survivor* in 1997, that more Spaniards watched *Big Brother* in 2000 than tuned in to Real Madrid's Champions League semi-final, and that in 2003 Norway's *Pop Idol* pulled 3.3 million SMS votes from a population of 4.3 million people.

The X Factor advances an alternative to democracy, the enigmatic 'X' of celebrity which replaces the ostensibly obsolescent cross on the ballot sheet. It seems no coincidence that when, in March 2010, in the run-up to the UK's 2010 election, BBC television launched a youth version of their flagship political discussion programme *Question Time*, they chose *X Factor* host Dermot O'Leary to front the show. Helen Wilkinson (2010: 47) has proposed that 'we have not yet arrived at the political *X Factor*, though we may not be far off.' Other commentators have suggested it is already here. A month before Britain's 2010 election, in an episode of the BBC situation comedy series *Outnumbered* a small child gives a description of how she imagines the British election process works – it is remarkably similar to the workings of reality television: 'Is there like lots of people and then they say "the lines are now open" – and then you vote off all the annoying ones until there's just one left and then they go – "I'm so happy, I'm Prime Minister now"?' Her father responds that she is confusing democracy with *The X Factor*. The implication, of course, is that this is what we are all doing. Indeed the columnist Johann Hari published a piece in *The Independent* in April 2011 arguing in favour of the Alternative Vote electoral system (in the weeks before a UK referendum failed to introduce such a system) explicitly on the grounds that it mirrored the voting structure of *The X Factor*.

The BBC's political correspondent Ben Wright pointed out at the end of April 2010 that the then ongoing general election was 'a TV election – almost like *The X Factor*.' Two weeks earlier the BBC's political discussion programme *This Week* had opened with an extraordinary song and dance tribute to *The Wizard of Oz* – an apparent attempt to combine their ongoing election coverage with a popular reality television series then airing entitled *Over the Rainbow* (in which musical theatre legend Andrew Lloyd Webber sought to cast Dorothy in a West End production of that show): the programme went on to feature a star of Sky's reality television series *Pineapple Dance Studios* offering advice to the UK's three main political party leaders on how to present themselves in the televised prime ministerial debates. *The Guardian*'s television critic and columnist Charlie Brooker, speaking on Channel 4's *Alternative Election Night* coverage on 6 May 2010, described the ground-breaking televised debates between the leaders of the UK's three main political parties which had aired during that campaign as 'a political version of *The X Factor*.' (Conversely, Brooker has elsewhere depicted the reality television series *America's Next Top Model* as a 'general election for people who'd vote for sequins.')

On 24 April 2010 – at the height of a British general election campaign – the BBC's *Breakfast* news programme cited a survey of young children which suggested that many would like to see Simon Cowell elected Prime Minister. The significance of such reality television series to the British political process was underlined by *The Sun* newspaper's front page head-line the morning after the election of spring 2010: 'Cameron wins the eXit factor'. Indeed (although its later edition ran with 'Election blues') *The Sun*'s News International stablemate *The Times* had led the first version of its front page that morning with a headline which not only stressed the unknowability of the election result but also referenced the reality-television-style atmosphere of the contest: 'The X Factor'.

On 14 December 2009 the BBC had reported an idea advanced by *X Factor* creator Simon Cowell to further blur the boundaries between political reality and the escapist fantasy realm which we call reality televi-sion – an idea which, although unrealized for the election of spring 2010, offers a hauntingly possible projection of twenty-first century democracy:

> Cowell has designs on the next UK general election [...] a series of big prime-time shows leading up to the election in which the public would hear two sides of the argument about several issues. There would, he said, be a red telephone for the poli-ticians to ring in, a massive *X Factor*-style studio audience split for and against the issue, and live voting by the viewers.

In the run-up to the UK's general election in 2005, ITV had launched a reality show called *Vote for Me*, designed to allow the public to select a candidate for parliament. However, the programme was consigned to a late-night 'graveyard' viewing slot. One suspects, however, that if Simon Cowell had in fact launched his prospective political reality show, it would have been afforded a rather more prestigious timeslot, and would have given the viewer a greater sense of the significance of the event in which they were participating, a sense that their involvement was an instance of political activism in itself. Cowell told the BBC that his idea might include debates of such issues as capital punishment: one might imagine that if his audience had voted overwhelmingly in favour of the return of hanging, this would have put an unprecedented pressure on the incom-ing coalition government in a way that might be seen as undermining the foundations of representative democracy. It was nevertheless reported by the BBC that a spokesperson for the then Prime Minister Gordon Brown had responded to Cowell's idea that 'he welcomed attempts to promote democracy.'

In July 2008 *The Times* newspaper reported that Simon Cowell had been brought in by the British Conservative Party to help them 'showcase Tories' talents.' In December 2009 *The Guardian* added that Conservative Party leader David Cameron had suggested that 'there is probably something we can learn in politics' from Simon Cowell. The day before the general election Cowell was afforded *The Sun* newsaper's front page when he formally announced his support for Cameron's Conservative Party – and while he emphasized that he did not believe a general election was equivalent to *The X Factor*, one might therefore wonder why he thought his judgment on the candidates should be so prominently delivered to the readership of the UK's highest selling daily newspaper.

According (among other sources) to the *Daily Mirror*, *Metro* and the *Daily Record*, it was believed that Simon Cowell would receive a knighthood in the Queen's Birthday Honours List the following month. It was claimed that the recommendation for this honour was one of Gordon Brown's last acts as Prime Minister. However, although this had been widely reported, Mr Cowell's name did not appear on the subsequent Honours List, published on 12 June 2010. It appears this is one case in which the media hype did not entirely hypermediate political reality.

On 1 January 2010 *The Guardian*'s Marina Hyde had critiqued the Conservative Party's plans to adopt the demagogic strategies of Simon Cowell by launching plans for an online platform designed to allow the online public to 'resolve difficult policy challenges.' Hyde wrote:

> The Tories have solved the problem of their lack of policies: they are going to wait for the Internet to tell them what to do [...] I suspect we are seeing the first instance of Cameron's excruciatingly wrongheaded plan to plunder the oeuvre of Simon Cowell. First TV, now politics – the illusion of deferral to the crowd is the mania of the age [...] Once, there were big ideas in television programming and in government [...] These days, such risks have been jettisoned in favour of allowing the public to think they are writing the script.

Hyde identifies a crucial similarity between the persuasive strategies of reality television and those of such new media forms as the Internet – the illusions of agency afforded to their audiences. Her assertion that these same strategies have now been adopted by political parties is both convincing and portentous.

Cowell's support of the Conservative Party was echoed by Andrew Lloyd Webber's endorsement of the Conservatives and Alan Sugar's work on behalf of the Labour administration. In June 2009 it was announced that Sugar would be elevated to the House of Lords and join Gordon Brown's government in the position of 'enterprise tsar'. Sugar, like Lloyd Webber and Cowell, is not only a successful businessman – he also fronts his own reality television series, the BBC's *The Apprentice* (2005–). Because of Lord Sugar's political associations, the BBC chose to delay the screening of the sixth series of *The Apprentice* until after the election of May 2010; on 11 May 2010, however, *BBC News* reported Sugar's frustration that the reality TV series featuring his fellow peer Andrew Lloyd Webber, *Over the Rainbow*, had been screened through the election period: 'We saw him in *The Sun* newspaper last week saying he backs Mr Cameron.' Indeed on an edition of *Over the Rainbow* just over a week after the election Lloyd Webber drew parallels between that talent contest and the democratic process: 'each one of these girls has tasted rejection – a bit like our political parties.' While optimists might suggest that this blurring of distinctions between reality television and democratic processes represents a healthy dissemination and integration of the political within popular culture, others might suppose that this phenomenon results in a dilution and trivialization of the ideals and substantive issues of politics which threatens to remodel democracy into the demagoguery of an ephemeral and superficial popularity contest.

Politics is thus becoming indistinguishable from reality television – and not only in the United Kingdom. In a television news report of 27 March 2010 the BBC's North America editor Mark Mardell described U.S. politics as increasingly resembling the extremes of reality television. As Barack Obama commented – during an interview on Jay Leno's talk show in March 2009 – 'Washington is a little bit like *American Idol*, except everyone is like Simon Cowell.' Obama's joke rebounded on him: the following week on the same chat show, Cowell suggested that the U.S. President had been upset when the pop guru had subsequently snubbed a dinner invitation: 'I was invited to have dinner with him last week, but wasn't available.'

Cowell made it quite clear who in his opinion wears the trousers in the relationship between politics and the entertainment media.

While *The X Factor* offers members of the public the opportunity to transform fellow citizens into celebrities, its network stablemate *I'm a Celebrity ... Get Me Out of Here!* (2002–) apparently sponsors the opposite process: the possibility of diminishing the status of the celebrity, phone-voting to force them, for example, to eat insects, grubs and spiders in the programme's 'Bushtucker Trial'. Yet the pleasures generated by both series are equally quasi-egalitarian – and may similarly stand in lieu of real world egalitarian or democratic impulses. This participatory illusion represents the defining myth of media-democracy, the cybertopian dream that the hegemony of the media offers an interactive process of popular empowerment. Yet, in the end, is the citizen of a mass-mediated world an active participant in a technologically enhanced democracy or the passive dupe of a mediacentric autocracy? Nick Couldry (2010) has suggested that reality television may increasingly be seen as becoming 'a wholly unacceptable form of social management' – and this perspective appears to be becoming increasingly prevalent in academic circles.

Can popular resistance to the hegemony of reality television then represent an upsurge in a desire for socio-political agency? In December 2009 a successful Facebook-based campaign to ensure that the *X Factor* winner's single failed to top the UK's popular music charts at Christmas did not, in the end, represent a resurgence of democracy so much as a consolidation of the process whereby popular action has shifted away from the realm of substantive political participation. This, then, was the extent of the triumph of the protest against globalized capitalism at the end of the week in which the Copenhagen summit on climate change failed to reach consensus upon a protocol which might adequately address an impending environmental catastrophe. The campaign to secure the top spot in the Christmas charts for Rage against the Machine's 'Killing in the Name' gave its participants an illusion of agency (of their resistance to power) which thereby subsumed any possibility for actual empowerment. The same might be considered true – even more so in terms of the pointlessness of the enterprise – in the case of the December 2010 campaign to beat the *X Factor* winner's single to the top of the charts with a version of John Cage's 4'33" (known as the

Cage against the Machine project) – in which dozens of musicians kept
silent for the prescribed period of time.

In his discussion of the anthropology of YouTube, 'The Machine is
Us/ing Us', the digital ethnographer Michael Wesch has suggested that the
global spread of YouTube users uploading their webcam videos of themselves dancing to a Moldovan pop song of 2003 might be described as 'a
celebration of new forms of empowerment.' If this is a mode of empowerment, then it is one which seems unrelated to conventional forms of
political power. It might of course be argued that truly radical empowerment would not be constrained by the traditional definitions or frames
of politics. It might also be pointed out that this is an argument which
serves the interests of entrenched power. If the disempowered populace
believe that messing around on Facebook or YouTube represents a real
form of empowerment, then they will not attempt to seize actual political power. All the better, then, for the traditional institutions of power.
The bread and circuses of contemporary media forms afford more than
vicarious pleasures; they offer their publics the impression of real agency,
a seductive illusion of empowerment whose dissolution its sufferers (or
addicts) relentlessly resist.

Yet perhaps there is no possibility of empowerment, only the fact
of power; no agency, only the illusion of agency. There is no functional
intentionality; there is only systemic evolution. We are not free agents; we
are merely reagents – objects sustained by an illusion of subjectivity. We
could call this illusion a displacement activity, if it were an activity at all;
we might instead call it a displacement illusion-of-activity. The illusion of
interactivity satisfies and therefore sublimates and dilutes the desire for
real active socio-political participation: we do not seek empowerment by
resisting external power because we believe we are already empowered.
The illusion of agency fostered by reality television, as by the video game,
recruits its subject into the ideological hegemony. The processes of interpellation (and ideological recruitment) fostered by these illusions of agency
are sustained in contemporary popular media forms by the appearance
of interactivity, the transfiguration of ordinary people into power figures
(celebrities or avatars), and impressions of democratic participation and
historical significance.

You've Been Framed

The paradox of the bourgeois-proletarian relationship is clear enough to the classical Marxist. What has been less clear is why this exploitative relationship continues. Why has the proletariat not cast off the shackles of its exploitation? Why has the revolution simply not taken place?

The apparently collaborative suffering of the proletariat perhaps to some extent recalls that of the *Sonderkommando*, the Jewish volunteers who, in Auschwitz, aided in the slaughter of their own people – and of each other (for the *Sonderkommando* were not excluded from the death lists). Robert Lifton and Eric Markusen (1991: 237) explain that the duties of the *Sonderkommando* included conducting their fellow Jews to the selections and then to the gas chambers. The function of this collaboration was, for the Reich, as psychological and ideological as it was practical: it involved a confusion of the actual monstrosity of the persecuting subject with the imagined monstrosity of the persecuted object, and therefore propagated a justification of sorts, a spurious claim of mitigating circumstances (if these people can do this even to themselves, they truly deserve to die). It was as if, as Primo Levi (1989: 37) wrote, the architects of genocide were announcing to their victims: 'we can destroy not only your bodies but also your souls, just as we have ours.' The dehumanization of the victims – their transformation, actual or imagined, into beasts or machines incapable of human feeling – was for the Third Reich, as for George Orwell's Big Brother, the ultimate abdication of moral commonality, and therefore of moral responsibility.

Why then do the 'victims' of power submit to – even or collaborate in – the processes of their own victimization? Is it because it is psychologically less agonizing to abdicate one's individuality, and to be assimilated within the dominant ideology, its myths and its discourse – even if that system is designed to destroy you?

As Michel Foucault (1991: 26) points out, 'power is exercised rather than possessed.' Power structurations are self-performing and self-perpetuating; societal systematization is determined not by the conspiracies of sharp-suited men in smoke-filled rooms but by the evolution of institutional, economic and ideological conditions. These structurations are, in Bourdieu's terms (1977, 1986), collectively and objectively orchestrated by institutional systems rather than by individuals. In these terms, we are no more than the vehicles, vessels or tools of Marx's ideologies or of Richard Dawkins's memes: as Marshall McLuhan (2001: 51) supposed, humanity becomes no more than 'the sex organs of the machine world, as the bee of the plant world, enabling it to fecundate.'

It may be suggested that the persecuted and exploited (such as the Marxian proletariat) embrace the system of exploitation because they believe that one day they will become the exploiters themselves – that one day they will win the lottery that is capitalism. Yet this theory does not in itself explain why the proletariat should believe themselves capable or deserving of such a socio-economic transfiguration. The reason that we each think that we will one day beat the system (or, rather, win within the system) is perhaps that we have been programmed to think of ourselves in this way. During the 1990s there was a much-screened television commercial for Britain's national lottery in which a giant golden hand (the hand of God?) appeared from the heavens and pointed its index finger towards a person who had just bought a lottery ticket: 'It could be you!' We buy lottery tickets because we believe that (against all the odds, both astronomical and divine) there is something special about ourselves which means that we will win. Perhaps we collaborate in power systems which exploit and persecute us for similar reasons. Why then do I think *it could be me*? That, argued Louis Althusser, is a matter of *interpellation*. It is, as L'Oréal would say (in one of advertising's most spectacularly overt moments of interpellation), because you're worth it. For Louis Althusser, mass culture calls to us in precisely this fashion; it announces to us that our lives are meaningful, that we are each of us at the centre of the universe. It is only this belief which allows us to serve the state as sane, submissive, functional citizens. We recognize ourselves in our media heroes, and therefore dream of our own meaningfulness. Just as Clarence the angel creates for James Stewart a fantasy of individual meaningfulness in Frank Capra's *It's a Wonderful Life*, so the process of interpellation posits the passive individual as an active subject: the individual is thus assimilated within the dominant structure of power.

In November 2009 LG Electronics launched an advertising campaign for its range of LED televisions which announced that 'the day we are born is the last day we are truly free' – but proposed that its 'seamless entertainment' hardware could somehow deliver us from this existential trap. As *Marketing Direct* noted, the commercial suggested that 'we are "boxed in" by life and jobs. With wireless technologies, the ads claim we can be "free" again.' This vision offers a release from the ties that bind us into the banal absurdity of post-industrial existence – through the natural and seamless logic of digital entertainment, one whose invisible sutures deny and therefore disseminate – as Silverman (1984) and Fiske (1987) would say – the ideological foundations of their own construction.

In 2010, Sweden's TV licensing body *Radiotjänst i Kiruna AB* launched an online viral campaign to promote the payment of its licence fee. This campaign involved a video in which a government official calls a press conference, with the eyes of the world upon her, to announce that a hero has come within our midst: 'This person gives us an alternative to uniformity. We owe this person for making an ordinary day into something special.' The official goes on to ask: 'How can you really trust that what we see on TV and hear on the radio is true? How can we be sure that the weak voices are heard and not scared into silence?' Her answer is that it is of course this heroic individual who allows us to understand and to ensure 'that our opinions are really our own.' The face of this hero is shown through the video on placards and billboards and in art galleries: and the face is your own. (This is achieved simply by loading your photo onto the campaign's website.) By paying your television licence fee you have heroically assured the continuation of freedom, fairness and democracy.

This video takes Althusser's notion of *interpellation* absolutely literally: in order to achieve its ideological ends – in order to construct its viewer as a responsible citizen who pays her licence fee (as a literal subject of the mass media) – it demonstrates how one may, in doing so, become the central figure in one's own universe, how one may turn out, as Dickens's David Copperfield said, to be the hero of one's own life. This illusion of power is itself, of course, utterly disempowering.

This, then, is why the revolution will not be televised: because the interpellative processes of the mass media negate the very possibility of revolution. The revolution will not be televised, because the televisualization and televization of reality (its conversion into and dissemination as television) involves the very negation of popular political activity or participation. The illusion of interactivity satisfies (and therefore diminishes) our desire for participation and subjectivity: because we believe we are active participants, we no longer strive to be so; because we believe we are self-determining subjects, we allow ourselves to be transformed into ideological objects. In this context it appears that the mass media's much-vaunted role as the mediator between people and power may in fact disguise quite the opposite function, and that the more the media announce their potential for interactivity and dialogue, the more monological and hegemonic they may invisibly become.

Liesbet van Zoonen (2004: 20) has suggested that one of the key reasons for the popularity of reality television lies in the desire to legitimize one's private life as normative and authentic. This seems to be related to Althusser's notion of interpellation and to the reinforcement which Bourdieu (1977: 80) witnesses afforded by the individual's observation of similarity in the experience of others. Yet, of course, reality television only offers its publics a vicarious and illusory validation of their integrity as individuals: in fact, by making the private public, rather than recognizing the universality of the authentic and original individual, it homogenizes individual specificities and denudes privacy of its defining quality, the authenticity which acts as an exemplar for public conduct. Reality television offers the individual viewer the prospect of a transfiguration into an avatar which projects its own individuality onto the wider (public, civic, social and political) world but in effect this illusion of potent agency merely veils the transformation of the citizen into the media consumer. A parallel process of transformation appears also to be evident in the pervasive structures and practices of social networking online.

CHAPTER 5

Social Networks

In a tongue-in-cheek editorial piece for *The Journal of the Federation of American Societies for Experimental Biology*, that journal's editor-in-chief Gerald Weissmann (2009: 2017) imagines the inter-war German Marxist cultural theorist Walter Benjamin reincarnated today as the dedicated flâneur of the Internet: he 'ambles (*surfs*) along a protected space (*MySpace*) in which bustling crowds are reflected in shiny *Windows*. He adjusts his cravat in a store-front mirror (*Facebook*), and when the *bell-tone* rings in his pocket, he takes out his timepiece (*BlackBerry*). He looks past his mirror image (*YouTube*), to find two generations of *followers* (*Twitter*).' Although Weissmann suggests that Benjamin might relish new media culture, he also draws our attention to Benjamin's famous adage (from his *Theses on the Philosophy of History*) that every document of civilization is also a document of barbarism. For Benjamin, as we have already seen, the process of absorption in the work of culture may be viewed as offering the potential at once for political liberation and for ideological enslavement; and so the Internet itself may be seen simultaneously as a realm of information upon which we surf, self-determining and critically independent, or as one whose paradigms we absorb into our selves, whose structures overwhelm without challenge our ontological, societal and political subjectivities. Benjamin's work is suffused with an anxious ambivalence in relation to these processes of mediation. As such, his celebration of electronic interactivity might have been rather less enthusiastic than Weissmann suggests.

The Subject of Facebook

In September 2009 the BBC quoted Facebook's vice-president of engineering, Mike Schroepfer, on the company's mission 'to get as much of
the entire world on the social network.' Facebook's own statistics (as of
January 2012) suggest that 483 million users are active on Facebook on a
daily basis. In May 2012 it was reported that Facebook had accumulated a
grand total of 900 million registered users. Facebook's ambition is overtly
directed towards global domination and by implication the homogenization of the contexts and structures of social interaction. Indeed, in April
2010 *BBC News Interactive* reported on what it described as 'Facebook's
bid to rule the web.'

Facebook's own Facebook page announces that 'Facebook's mission is
to give people the power to share and make the world more open and connected.' It is unclear whether this strategy includes the use of a public relations company to plant negative stories about its rival Google, as reported
in May 2011. On his own Facebook page the site's founder Mark Zuckerberg
lists his personal interests as 'openness, making things that help people connect and share what's important to them, revolutions, information flow,
minimalism'. Facebook as such represents a revolutionarily minimalist
notion of information flow, one in which the flow itself, the process of
connection and of sharing, and the condition of openness which affords
that possibility, signify more, absolutely more, than the content itself, insofar as what is important to the Facebook user, as to its founder, are these
processes themselves. Facebook emphasizes the significance of the act of
expression rather than what is expressed. Despite its promises of interactivity, its own processes thereby define its users as the subjects of essentially
empty and banal monologues.

Petitioning its user for a status update, the empty field at the top of
one's Facebook homepage asks: 'What's on your mind?' Not *what are you
doing?* – the life of the user may have become too passive, too solipsistic, to
countenance such a call to action or to material interaction: the Facebook
user becomes a voice devoid of context, one which expresses itself and only
itself, a subjectivity defined by and equal to the processes of Facebook use.

In May 2010 Danny Sullivan wrote on the Search Engine Land blogsite that he had recently attempted a Google search on the phrase 'how do I' – only to discover that '"How do I delete my Facebook account" was one of the top choices.' Sullivan added that 'if you go back to Google and start typing in "del", you get "delete facebook account" as the top suggestion.' (Trying this in 2012, one actually has to type the first four letters before this suggestion appears.) In a subsequent blog entry that same month Sullivan suggested that, although the company was unwilling to provide him with its cancellation statistics, it appears that Facebook users may be cancelling their accounts because of concerns over the company's privacy policies and thereby the use to which users' data may be put. The same month, Facebook announced a simplification of its privacy settings, which, according to company founder Mark Zuckerberg, had 'gotten complex' for users. As the BBC reported at the time, this move followed 'a storm of protest from users over a series of changes on the site that left its members unsure about how public their information had become.' Indeed, as Papacharissi (2010: 46) has reminded us, Facebook had been obliged to suspend its terms of service in the face of criticisms relating to the potential for privacy violations. In August 2011, Facebook announced yet another revision of its privacy policy, described by *BBC News* as 'the latest in a long line of attempts by Facebook to streamline how members manage their personal information' in response to criticisms that the social network had appeared 'to bury privacy settings in obscure menus.'

It may then be that Facebook's vision of openness and connectedness could be interpreted in terms of a data-mining strategy which categorizes and commodifies its users as packages of commercially valuable information. Facebook's concept of openness has not extended – as one might expect of any transnational corporation – to the publication of contact details of the senior figures and spokespeople within its structure of governance on a corporate website: indeed even the physical location of the company's offices is shrouded in some ambiguity (while the company's headquarters appear to have remained in California its so-called 'international headquarters' are now sited in Dublin). Facebook tends to expect rather more openness of its users than it is willing to provide to them. As Hodgkinson (2008) has suggested, perhaps somewhat hyperbolically, 'Facebook pretends to be about freedom, but isn't it really more like an ideologically motivated virtual totalitarian regime?' Although not exactly totalitarian, Facebook's short-lived Beacon system, launched in 2007, might be seen as commercially totalizing. This system allowed the identification of Facebook users who had registered a liking for a product in the advertising of that product to their friends. As Emily Maitlis suggested in her 2011 BBC documentary *Inside Facebook*, such schemes have prompted commentators to ask how far the site can go 'without its users feeling exploited' – or at least feeling so exploited that they actually threaten to quit – or indeed, in the Beacon case, to sue.

Facebook's openness then appears to be a one-way system: Facebook is open to new users, but appears to be relatively closed to users who want to leave. There is in fact a group page on Facebook (established by a member who is 'not a part of the Facebook team') which advises users 'How to permanently delete your Facebook account'. It explains itself thus: 'Ever tried to leave Facebook and found out they only allow you to "deactivate" your account? All your personal data, including photos, interests, friends etc will still be saved indefinitely! You don't have to be a conspiracist to find this quite fishy (or simply annoying)!' The subject of quitting Facebook is indeed considered so significant that in February 2008 *The New York Times* ran a feature on the company's agreement to make it easier for users to delete their accounts after reports that 'some Facebook users who wished to close their accounts had been unable to do so, even after contacting Facebook's customer service representatives.' Facebook, it seems, is not unlike Afghanistan. It is much easier to get into it than to get out. Its mission of liberation even, as we shall see, attempts to maintain its hold over its dead. What announces itself as a site of openness and liberation may in fact do its best to constrain its users. Oymen Gur (2010) has suggested that, while social networking websites offer their users the illusion of personal liberation, they effectively seize control of their users' subjectivities: 'even though social networking sites seem transparent, users are still mediated through them – the more people are liberated with wider and more transparent networks, the more they are constrained.'

This sense of constraint – of inescapability – is not, of course, limited to Facebook. Indeed, a Dutch website called Web 2.0 Suicide Machine has been set up to allow people to disconnect from – and delete their presence within – their online social networks. Its founder told the BBC in September 2010 that his clients 'are basically getting their analogue life back.' However, a psychiatrist who (according to the BBC) 'treats patients who use the Internet excessively' warned that 'by disconnecting you are losing a significant relationship. Those 30 or 40 hours of time now have to be filled with real life.'

Our real lives are, clearly, increasingly defined by our online existences. The subject of Facebook, the site's mediated user, is delineated by Facebook parameters (gender, relationship status, political views, religious views, interests in movies, books, television and music) and by an ideology of online friendship (friendship as process, friendship as social capital, but not necessarily friendship as moral commonality). One's relationship status has been limited to: *single, in a relationship, engaged, married, it's complicated, in an open relationship*, or *widowed*. One has been permitted to be *looking for* a very limited (and limiting) range of things out of Facebook use (and out of social life: which for some *is* Facebook use): *friendship, dating, a relationship* or *networking*. These limits are in themselves both psychologically and ideologically defining. Boyd and Heer (2006) have described the embodied interaction which takes place through social networking sites as a process through which individuals write themselves into being. Siibak (2009) argues that selfhood is constructed in social networking sites according to values associated with ideals or expectations of selfhood. Indeed, we might see those expectations as increasingly determined by the discourse structures promulgated by these sites themselves.

We are spending more and more of our time online, especially on social networking sites. In May 2010 a survey conducted by UK Online Measurement suggested that British web users were spending 65 per cent more time online than they had just three years earlier. The survey added that the largest single portion of that time (22.7 per cent) was spent on social networking or blogging sites (as opposed, say, to e-mail use which came in second at only 7.2 per cent of users' time). In November 2009, the BBC reported that a British schools organization had listed SNS addiction as the primary cause for parents' concern about their children's welfare, suggesting that some children appeared to have become 'permanently connected' to such sites as Facebook. It is certainly clear that we are spending increasing amounts of our time on social networking sites and that users' dependence on such sites is therefore increasingly influential in their definitions of selfhood. Within a few months of its launch on 23 September 2009 a Facebook page entitled 'I will name my son Batman if this page gets to 500,000 Fans' had garnered nearly a million supporters – resulting in the announcement that the unfortunate offspring of the page's owner had indeed been named 'Batman'.

Sites like Facebook offer their users a mode by which to 'capture and share [their] life story' (Kelsey 2010: 1). Updates to the site's structure led the *BBC News* to suppose in September 2011 that 'identities will now be defined through a densely packed vertical timeline of major life events.' Indeed the BBC quoted technology research analyst Sean Corcoran as supposing that 'Facebook is positioning itself as your life online.' Eager to report the most enthralling status updates, it is not perhaps uncommon for Facebook users, like diarists or autobiographers, to lead their lives and frame their identities according to the needs of their narratives. Paul de Man (1984: 69) wrote that 'we assume that life *produces* the autobiography as an act produces its consequences, but can we not suggest, with equal justice, that the autobiographical project may itself produce and determine the life and that whatever the writer *does* is in fact governed by the technical demands of self-portraiture and thus determined, in all its aspects, by the resources of his medium?' Roland Barthes (1977: 144) suggested of Marcel Proust that 'he made of his very life a work for which his own book was the model' and the users/content-generators of social media may be performing a similar practice. Although the popular video-sharing website YouTube may have adopted the slogan *Broadcast Yourself*, it is clear that many of the site's users are intent upon inventing new selves for themselves – or even, as in the case of such YouTube stars as EmoKid21Ohio or Lonelygirl15 (both videobloggers famously exposed as fictitious characters), upon counterfeiting such selves. These extreme attempts to forge one's own identity offer fitting metaphors for the processes of self-invention that to some extent underpin all acts of autobiography.

Samuel Johnson's advice to the prospective diarist (Johnson 1951: 176) remains timelessly appropriate when he cautions against 'an intemperate attention to slight circumstances [...] lest a great part of life be spent in writing the history of the rest.' One could do worse than to heed the wisdom of Richard Ayoade's character Maurice Moss in the 'Friendface' episode of Graham Linehan's situation comedy *The IT Crowd* (2008) as he explains his resistance to participation in social networking sites: 'I think I've got better things to do than talk to friends and flirt with people, thank you very much.'

Status Symbol, Status Update

Why might Facebook users so relentlessly endeavour to lead lives worthy of their status updates? Why might they interrupt the flow of their real-world interactions in order to post such updates onto the site? Although this activity may in part relate to an internal process of self-narrativization – a coming to subjective integrity through the reconstruction of one's life within the logic and aesthetic of the narrative form – it also represents an outward-facing expression of one's own public identity and social status.

While the status update announces and performs the identity of the user, it is specifically the user's number of Facebook friends which provides material evidence for the success and status of that user's constructed identity. Danah Boyd (2006: 13) supposes that one's collection of Facebook friends affords a space for identity performance. While Boyd (2008b: 213) has specifically argued that 'the public display of connections that takes place in social network sites can represent a teen's social identity and status', it seems apparent that this need for validation is not unique to Facebook's teenaged users.

Boyd (2006: 4) suggests that 'the public nature of [social networking] sites requires participants to perform their relationship to others [...] participants override the term "Friend" to make room for a variety of different relationships so that they may properly *show face*.' In other words, this process of identity construction, performance, affirmation and validation, insofar as its success is measured by the numerical accumulation of friends as a valorizing commodity, redefines the nature of friendship itself in order to accommodate the relational inflation which this global and virtual medium promotes. It used to be the case that one needed tens of friends to prove one's social and personal worth (to oneself as to others); within the psychological economy of Facebook one needs hundreds ... thousands. (And on Twitter hundreds of thousands.)

Boyd (2006: 10) proposes that 'by having a loose definition of Friendship, it is easy to end up having hundreds of Friends [...] Because of how these sites function, there is no distinction between siblings, lovers, schoolmates, and strangers. They are all lumped under one category: Friends.' One might also advance the argument that it is only possible to accumulate these hundreds of friends by changing one's definition of friendship. The social networking site seems, as Papacharissi (2010: 55, 63) suggests, a social realm which, while more fluid than traditional modes of socialization, appears fragmented, superficial and commodifying. Although Facebook has, since Boyd ruminated on this issue, introduced two complementary categories – 'close friends' and 'acquaintances' – this formal categorization and hierarchization only adds to the sense of friendship as an economic commodity.

Matthew Tedesco (2010: 123) has argued that 'our ordinary understanding of friendship is of a relationship, where that relationship requires something of us. We spend time with our friends; we invest ourselves in our friends. But interactions with our Facebook friends usually aren't like this. In some cases, we don't interact with them at all, and when we do interact with them, our exchanges are often ephemeral and impersonal, lacking any of the investment that we find in our ordinary friendships outside of cyberspace.' To what extent then can we call this strange, new, virtual thing by that old name of friendship?

The Geography of Friendship

Samuel Johnson's definition of the word *intimate* in his *Dictionary* of 1755 proposes, amongst other things, that the intimate is 'near; not kept at a distance.' Robin Dunbar (2010) has argued that, although social networking sites may have value in supporting ongoing offline relationships, relationships conducted entirely via such sites are neither as stable nor as strong as relationships which take place in the material world 'because you're not interacting face to face.' He has added that 'until they invent virtual touch you will never have networked relationships that are in any way like the ones you have in real life.' There therefore appears to be a real correlation between physical and emotional closeness. As Aristotle (2004: 210), one of western civilization's first theorists of the nature of friendship, argued, friends who do not spend much of their time together are not really friends.

Meditating on Aristotle's position, Jacques Derrida (2005: 222) has argued that 'if absence and remoteness do not destroy friendship, they attenuate or exhaust it.' Derrida goes on to point out that Aristotle's argument did not allow for the possibility of friendship mediated by telecommunications. It was not that Aristotle ignored the possibility that friendships might be sustained through contact via media of telecommunications because such media did not exist in his time (they did: there were, for example, such things as letters); Derrida implies that he ignored these possibilities because he did not consider them a viable way to maintain friendships.

New technologies have not abolished distance, although they can allow us to forget its enduring significance. When we interrogate the notion of online relationships, geography therefore seems both a banal and an essential factor in the ongoing shift in the concept of friendship. The moral – even, for some, spiritual – aspects of traditional notions of social and emotional intimacy seem paradoxically dependent upon the material conditions of that intimacy. From this perspective the limits to the possibilities of virtual friendship therefore appear insurmountable. Craig Condella (2010: 114) has asked whether 'the friendship spoken of by Aristotle is altogether different from the virtual friendships formed on sites like Facebook.' He returns in this context to the Aristotelian problem of (and necessity for) physical presence and physical action: and this, he argues, is where Facebook cannot deliver.

The Nature of Facebook Friendship

Boyd (2008b: 211–212) writes that friends on social networking sites are not necessarily equivalent to friends in physical life: 'participants use the term "Friends" to label all connections, regardless of intensity or type.' Boyd (2008a: 17) proposes that social networking sites may foster a 'fake sense of intimacy' with people one does not really know. Boyd (2006: 10) therefore suggests that, within the social networking site, 'there seems to be little incentive to be selective about Friendship [...] As far as most participants are concerned, Friendship doesn't mean anything really.' One might therefore ask whether the non-selective view of online friendship (if friendship doesn't mean anything) is changing traditional views of friendship in offline interaction – or whether SNS users are able consciously to distinguish between these modes of online and offline relationships?

Danah Boyd has responded that 'fundamentally people know the difference between a Friend (as in social network sites) and a friend (as in the intimate connection). This is why I go out of my way to differentiate them.' The obvious question is why one would need to go out of one's way to differentiate these modes of friendship if these distinctions were truly so clear. The globalizing structures of such social networking sites as Facebook are imposing a homogeneous cross-cultural notion of friendship upon the world; and insofar as users increasingly interact in a primarily online environment, their ability to distinguish between online and offline modes of friendship seems less and less relevant. As the phrase 'social networking' has come almost exclusively to refer to *online* social networking, is it so difficult to imagine a time when the word 'friend' might refer primarily to online friendship? Indeed, when Boyd (and others) view the Facebook friend as a 'Friend' (with a capitalized 'F') in an attempt to differentiate this phenomenon from offline manifestations of friendship, this translation of the term into a proper noun may suggest there is something absolute, definitive or seminal about this mode of friendship (in parallel, say, to the differentiation of 'God' from 'god').

According to Facebook's own statistics, the average user spends more than 55 minutes on the site every day. This suggests that, even if Facebook users remain able to distinguish between modes of online and offline friendship, many are committing significant social investment to their online relationships. Indeed, a study reported in *The Daily Telegraph* in November 2010 announced that one in four Britons socializes more online than in person. Smallwood (2010) has meanwhile reported the extreme example of one Facebook-addicted family in which children no longer made eye contact or talked with each other. One is tempted to imagine that if the Internet were ever irreparably broken (admittedly an unlikely scenario) there might be many young people who no longer had the social skills needed to go out and meet people in physical environments: one can envisage, in this absurd scenario, youngsters sitting in bars handing each other notes revealing their status updates and requesting friendship.

A member of Facebook's Data Team, Adam Kramer, has, however, commented that:

> It's important to remember that the vast majority of friendships on Facebook are actually traditional relationships that happen to also have a Facebook representation which is termed 'friend'. As far as friendship-like social dynamics are concerned, there's a decent amount of empirical evidence suggesting that communication through Facebook serves to strengthen 'weak-tie' bonds, or relationships among people who are not very close. But most broadly, it is my belief that Facebook and social networking sites are a facilitator of social interactions, and do not indicate anything 'new' about the psychology of human interaction. Rather, they serve as tools to encourage social interaction in new ways via new means of communication, much like the printing press, telephone and television did in their day.

Yet, without being overtly technologically determinist, one might suggest that Gutenberg, Bell and Baird's inventions did more than facilitate and accelerate communication; one need not be a disciple of Marshall McLuhan to imagine that these media technologies radically affected the nature and content of social interaction.

When new media technologies (such as the Internet and the mobile telephone) converge, one would expect their impact to be all the more intense. Facebook statistics reveal that more than 100 million users access Facebook through mobile devices, and that those who do so are twice as active as other Facebook users. A report from the UK's Office of Communications, published in August 2010, showed that Facebook use accounted for 45 per cent of total time spent online on mobile phones. The engineer who designed the hugely successful 'Facebook for iPhone' package, Joe Hewitt has commented that:

> Facebook hasn't affected contemporary notions of friendship significantly. Unlike some other social networks where one is encouraged to 'friend' people they've never met, Facebook works best when it is a mirror of the real world, and your friends are people you know in real life. The biggest effect Facebook has had on contemporary friendship is merely to keep friendships alive that might have otherwise faded due to time and distance.

Both Kramer and Hewitt see Facebook as merely a facilitator for traditional offline friendships. The medium is, it seems, no longer the message; the nature, content and meaning of a communication act are no longer affected by that act's own conditions of process and possibility. Burke, Marlow and Lento (2010) have similarly supposed that social networking sites merely complement real world relationships. However, Moira Burke has more recently added:

> Social networking sites certainly changed the threshold and nuance that most people experienced in the face-to-face definition of 'friend'. This is primarily because they force people to make relationships explicit, and 'friend' status was often simply an access control mechanism for content. Since early SNSs didn't have tiered relationships or faceted privacy controls, all kinds of social roles were collapsed into a single term. Between the awkwardness of having to explicitly deny someone's friend request and the publication of one's friend count on the profile, many users hoarded friends, in a socio-economic way.

In Judd Apatow's 2009 film *Funny People* Adam Sandler comments that the more friends you have online, the fewer friends you have in real life. One might at least suggest that one's focus upon the modes of friendship promoted and defined by social networking sites may change the nature of one's relationships offline, and that the amount of time spent involved in these online processes of social interaction, the number of 'friends' one has online and the emphasis one places not only upon these modes of interaction but also upon the significance of the quantities of friendships achieved and maintained online may radically affect one's notion of friendship itself.

In a reflection upon the congruence between the experience of Facebook and the playing of video games, Robin Hunicke (2008) has suggested that Facebook allows its users to feel that they are living enjoyable lives and that they are loved: 'Facebook makes people feel like they matter.' As such, one might argue that it offers not the liberating pursuit of happiness but the sentimental impression that one might already be happy and that this is perhaps the extent of what happiness is. This is, of course, yet another illusion of empowerment which pre-empts and prevents the struggle for actual social, political and cultural power.

In a Facebook blog entry of October 2009 Adam Kramer described a process whereby Facebook's team of data scientists have sought to quantify the sum total of American Facebook users' happiness: 'Every day, through Facebook status updates, people share how they feel with those who matter most in their lives. These updates are tiny windows into how people are doing. Grouped together, these updates are indicative of how we are collectively feeling.' Kramer went on to explain how 'data scientists at Facebook started a project to measure the overall mood of people from the United States on Facebook, based on the sentiment expressed in status updates.' In March 2010 *The Sun* newspaper reported that, according to a parallel survey conducted by Facebook in the UK, one of the happiest days of 2009 was 13 December – the day of the season finale of the reality television series *The X Factor*. It seems no coincidence that Facebook users find feelings of happiness (a sense of social and emotional inclusion) through reality television, a media genre whose strategies echo the illusions of participation promoted by so many interactive media technologies.

The Commodification of Friendship

An exercise has been conducted to test the extent to which Facebook users exercise discrimination in their acceptance of friends. A dummy Facebook account was created under a name which, although ordinary and even bland (by English-language standards), did not exist on Facebook – nor (according to a Google search) anywhere on the Internet. This name was chosen specifically to avoid the possibility that it might be interpreted as a real known friend. The Facebook account did not include any images of the account-holder nor any profile information. In stage one of the exercise 100 Facebook users were sent friend requests: the names of these users were generated randomly. Within five days 18 per cent of these Facebook users had accepted the dummy user as a friend.

In the second stage of this exercise a further 100 friend requests were sent to Facebook users. These recipients were all contacts generated by Facebook itself as suggested friends – in other words they were friends of those who had already accepted the dummy user as a friend. Within five days 74 of these friends of friends had befriended the dummy user.

In the third stage of this exercise a further 100 friend requests were sent to Facebook friends of those who had accepted Facebook friendship in stage two (i.e. friends of friends of friends). Within five days 84 of these friends of friends of friends had befriended the dummy user (see Figure 1).

These results appear to demonstrate that while the majority of Facebook users are relatively discriminating in accepting the friendship of completely unknown contacts, the majority are also willing to accept such friendship on the implicit recommendation of one of their current friends; that is, if they know someone who knows the person requesting friendship, they are likely to accept that friendship. Indeed, they appear to be even more likely to accept that friendship if more than one friend is already shared, and if the petitioner already possesses a significant number of friends. It appears that any friend of anyone who is already a friend of mine is also a friend of mine. This mode of friendship is ostensibly one based entirely upon quantifiable notions of social capital: the quantity of social

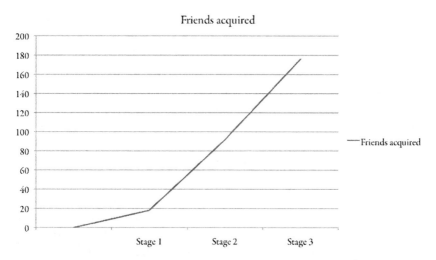

Figure 1. Numbers of friends acquired in three-stage Facebook friending exercise.

capital possessed by the seeker of friendship (the number of shared friends or their total number of friends) is directly related to the probability of the success of their petition. The petitioner's perceived level of social capital may therefore be seen as adding to the social capital accumulated by the petitioned party. This inference is further evidenced by the fact that during the second and third stages of this exercise the dummy account received seven unsolicited requests for friendship from friends of friends who had not been contacted – as this non-existent user accumulated social capital he came to be perceived by third parties as a commodity worth investing in. In Aristotle's terms this is therefore a mode of friendship based upon utility rather than upon virtue or moral commonality or proximity.

We might therefore ask whether social networking sites' utilitarian commodification of friendship in terms of the quantity of friends accumulated and that quantity's relationship to social capital has affected the popular concept of friendship. Has contemporary friendship become more a matter of socio-economic functionality than of emotional or moral idealism – or is this in truth what friendship has always been about?

The importance of numbers in terms of the functioning and value of Facebook friendship is perhaps most cynically demonstrated by the case of Sanford Wallace who in August 2011 was charged with sending more than 27 million spam messages to Facebook users. Mr Wallace was accused of developing a program which posted messages on Facebook users' walls – purportedly from friends – encouraging them to visit an external website which then harvested their personal details. This case demonstrated not only the risks of the anonymity and unaccountability of Facebook friendship, but also underlined the economic significance of the commodification of such relationships.

Burke et al. (2010) have argued that the uses of social networking sites (and specifically friend counts) have been associated with increased levels of social capital. Ellison et al. (2007: 1161) have also noted a correlation between Facebook use and the accumulation of social capital (cf. Valenzuela et al. 2009; Zywica and Danowski 2008) – while Papacharissi (2010: 139) has noted more generally that online communication has been associated with the creation of social capital.

This utilitarian commodification of friendship was depicted by Facebook's founder Mark Zuckerberg in April 2010 as the website's primary function when he spoke in somewhat enigmatic terms of Facebook's capacity to allow its users to 'use' their friends: 'If you look back a few years ago and even as recently as today, in most cases the web isn't designed to use your friends. We want to be one of the things that empowers that and right now most users are using Facebook.'

If friends are a commodity whose quantity enhances one's level of happiness, they are clearly one to be valued, and one whose integrity is to be regulated by the appropriate authorities. In November 2009 it was reported that Facebook was threatening legal action against a company called USocial that offered to sell its users' friends and followers on Facebook and Twitter. (On Twitter, for example, 1,000 followers cost £53.) In May 2012 it was reported that Facebook had introduced a system whereby users could pay a fee to increase the prominence of their posts. Friendship has clearly become an economic commodity whose quantification is a crucial concern. In a Facebook blog entry of April 2009 Facebook's Chief Operating Officer, Sheryl Sandberg wrote:

One of the most common questions we're asked at Facebook is, 'How many friends can you have?' It's an increasingly important question as more people around the world share and connect on Facebook and on the Web overall, but it's also difficult to answer. While the average user on Facebook has 120 confirmed friend connections, that number doesn't account for all the different types of relationships people have in their lives [...] When our Data Team measured active networks for users on Facebook, it found that, in any given month, users keep up with between 2 times and 4 times more people than through more traditional communication.

A technology which can increase one's quantities of friendship – and therefore one's levels of happiness – fourfold is clearly one which is to be cherished. A number of key questions however arise. When one increases one's number of friendships fourfold is each of those friendships as valuable as when one had fewer friends? Can friendship exist in a virtual environment which excludes the degree of material (spatial and temporal) commitment, dedication and even sacrifice which in the offline world tests, tempers and defines it? When friendship becomes a quantifiable commodity and a reserve of economic capital, is it still friendship? When we count our friends, do they count as friends? And, if not, might this increasingly bankrupt concept of online friendship return to undermine the integrity of friendship in the offline world?

Such newspaper stories as 'Facebook gatecrashers wreck family home at teen party' (*Daily Telegraph*, November 2008) and 'Family home trashed after Facebook party goes wrong' (*Daily Telegraph*, February 2010) or the tale of 'the teenagers who organised a Facebook party that descended into a street brawl' (*Daily Telegraph*, July 2009) or the account of 'the German teenager who advertised her 16th birthday party on Facebook only for 1600 gatecrashers to turn up at her house' (*Daily Telegraph*, June 2011) do little to promote a perception of the idealism of Facebook friendship. In August 2009, Vincent Nichols, head of the Roman Catholic Church in England and Wales, expressed his concern that sites such as Facebook 'encourage teenagers to view friendship as a commodity.' The Archbishop added: 'It's an all-or-nothing syndrome that you have to have in an attempt to shore up an identity; a collection of friends about whom you can talk and even boast. But friendship is not a commodity, friendship is something that is hard work and enduring when it's right.'

A similar phenomenon whereby one's number of friends or 'followers' determines one's social and cultural status may be witnessed in the Twitter platform. As Jerome Taylor wrote in *The Independent* newspaper in August 2010:

> You might be the 21st century's Oscar Wilde, with an acerbic wit of such magnitude that you can distil the world into razor-sharp aphorisms of just 140 characters. But unless you have people 'following' you, no one will even know you exist.

The quantification of friendship, or the delineation of its use-value, reimagines the ideal of friendship as a commodity. This commodification negates the very concept of friendship itself, one from which such seminal thinkers as Aristotle (2004: 204), Cicero (2010: 14) and Montaigne (1991: 207) specifically exclude those so-called friendships based upon individual utility or benefit. Of course such relationships exist and may be called friendships, but they do not represent that absolute Aristotelian ideal in which friendship is conceptually grounded. Utilitarian friendship represents a lesser, shallow or empty version of the ideal.

The Arithmetic of Friendship

The anthropologist Robin Dunbar has famously postulated what is known as Dunbar's number – the number of people with whom one can maintain social relationships. He has suggested that this number is about 150. Dunbar (2010) proposes that one can maintain close relationships with an inner core of only about five people, and that concentric sets of relationships demonstrating lesser levels of intimacy radiate from this core: 'as you go out you include more people but you're including relationships at a lower quality. That ties up very closely with the amount of time you spend with those people. The layers scale in very consistent patterns, they occur at 5, 15, 50, 150, 500, 1,500.' Indeed Dunbar (2010) extrapolates the next level in these relationship scales to agree with Plato that the ideal size for a democracy is about 5,300 people.

Dunbar (2010) has suggested that the accumulation of Facebook friends has become competitive, and argues that those with hundreds of such friends probably do not know most of them. This, he says, is both because of human cognitive limits and because it is not possible to invest sufficient time to sustain such intimacy with so many people.

Aristotle similarly suggested that the number of friends one can maintain is limited by the number of people with whom one can sustain intimacy, and therefore proposed that those who boast many friends are in fact 'friends of nobody.' (Aristotle 2004: 251). In his reading of Aristotle's notions of friendship, Jacques Derrida has argued that when you try to quantify friendship as an economic commodity (and therefore when you try to increase your number of friends in order to increase your social capital, the utility adhering to those friendships) then they are no longer friends in Aristotle's sense. Derrida (2005: 21) suggests that friendships are not objects or numbers, and that friendship itself resists quantification.

Derrida's analysis of the history of the concept of friendship in western civilization then turns its attention to the work of the sixteenth-century French essayist Michel de Montaigne to argue that this concept requires both intimacy and exclusivity. Montaigne (1991: 215) speaks of the purity of that friendship in which 'each gives himself so entirely to his friend that he has nothing left to share with another.' One of the most celebrated of friends in the history of letters, the lexicographer Samuel Johnson's long-suffering companion James Boswell (1986: 247) also saw friendship as discriminating and exclusive: he supposed that one must prefer the interests of a friend against those of others and endorsed what that 'old Greek said, "He that has *friends* has no *friend*."' (The old Greek, in question, was Aristotle, although it is not clear whether Aristotle quite said this – though Erasmus and Montaigne, amongst others, credit him with this adage.)

Can such intimacy, requiring such exclusivity, be possible in the resolutely and uncompromisingly public space of Facebook? Facebook's suspension of the contradictory imperatives of intimacy and openness (to some extent determined by the needs of its users and of its advertisers respectively) may not be sustainable. The dramatic dip in its share price in

the immediate wake of Facebook's first flotation in May 2012 may suggest
that such misgivings as to the sustainability of its model may be growing
in influence. The same month the BBC's Rory Cellan-Jones pointed out
that Facebook founder Mark Zuckerberg had not listed his wedding that
month as an upcoming event on his Facebook page and had therefore not
been bombarded by Facebook adverts from cakemakers, florists and bridal
couturiers. Cellan-Jones went on to ask whether therefore Facebook's model
of 'frictionless sharing' is sustainable when even its own founder seems
unwilling to share his personal details. As David Kirkpatrick (2011: 333)
has suggested 'reciprocal personal connections packed with very private
data may not coexist well with unbridled sharing.' Kirkpatrick goes on to
ask whether a system so deeply founded upon personal trust could ever
really be transparent and open. The ideology of Facebook – its mission
of unremitting, uncompromising and obligatory openness, its removal
of barriers, of distinction and discrimination – makes this seem unlikely.
When everyone is special, no one is.

Facebook friendship represents a relationship without responsibility
or commitment. In March 2010 *The Daily Telegraph* reported concerns
aired by Teesside's Director of Public Health Professor Peter Kelly that
Facebook use had contributed to a rise in sexually transmitted diseases, in
that such social networking sites were facilitating casual sexual encounters.
One might add that it is not merely the convenience of social networking
offered by Facebook but the casual illusion of emotional intimacy which
such sites promote which might provoke inappropriately prolific manifestations of physical intimacy.

Boyd (2006: 3) argues that 'the category of friend carries an aura of
exclusivity and intimacy.' Aristotle, Montaigne, Boswell and Derrida would
agree. A corollary of this proposal is evident: that the Facebook contact is,
within such traditional definitions, not a friend as such, not in the absolute and definitive sense. The meaning of friendship is therefore in the
process of changing: it is not only that the *concept* of Facebook friendship
(a new and radically different sense of friendship) is overwhelming that
of offline friendship, it also seems that the *practice* of offline friendship is
being overtaken by the practice of friendship online.

But does this matter? Is there any reason to cling onto the traditional notion of friendship? Beyond its moral and aesthetic idealization, does it have any meaning, value and function in itself?

The Politics of Facebook Friendship

Friendship has seemed essential, for several significant European thinkers, to the perpetuation of democracy itself. Aristotle (2004: 220), for example, supposes that 'while in tyrannies friendships [...] are little found, they are most commonly found in democracies because the citizens, being equal, have much in common.' Friendship is as such a prerequisite for, and a symptom of, democracy: as Derrida (2005: 22) points out, for Aristotle at least 'there is no democracy without the "community of friends".'

Papacharissi (2010) suggests that the maintenance of the private social sphere sustains the possibility of a broader socio-political public sphere (in that the public is founded upon the private) – but that, insofar as such new media forms as social networking sites blur the distinctions between the private and the public, the dissolution of the integrity of the private sphere also undermines the integrity of public interactions. Derrida (2005) argues that the concept of friendship lies at the core of the western political tradition, and specifically at the heart of the notion of democracy: democratic trust and interactivity is extrapolated from the ideal of friendship; and therefore Plato's ideal democracy grows out of Aristotle's ideal friendship. The absolute and irreducible discrimination of friendship stands as a benchmark by which to measure our broader social and political relationships.

If all the world were Facebook friends, would that then spell the end of 'real' friendship, of real enmity and even of democratic politics? This 'world without politics' – as Derrida (2005: 130) puts it – *already* appears to have been 'abandoned by its friends as well as its enemies' and seems little more than 'a dehumanized desert.' We return yet again, then, to Baudrillard's desert of the real – the waste land of the new, virtualized reality. Can social networking sites really hope to restore what has been lost?

The loss of the concept of friendship, and therefore of its essential and inevitable counterpart *enmity*, fosters a reality without political orientation. Derrida (2005: 77) laments the loss of these enemies without which we lose our political being. Derrida's study of *The Politics of Friendship* was written in the early 1990s, after the end of the Cold War and before the spectacular events of 9/11 and way before the launch of Facebook. Derrida (2005: 84) nevertheless, even then, came to suggest that the re-invention of the friend and of the enemy would be necessary to reverse this process of depoliticization.

We have, of course, reinvented the enemy: as Poole (2002: 2) points out, the new world order forged in the wake of 9/11 has invented an imaginary structure of allegiance, opposition and exclusion. Yet if the enmity, like the friendship, of contemporary western civilization is increasingly indiscriminate, abstract and virtual, then is the resultant politics anything more than an empty simulacrum? The 'big bad' of fundamentalist terrorism is no more real than the response it promotes – the response it was imagined in order to promote.

Stuart Price (2010: 176) argues that in the hectic attempt to justify the War on Terror too many enemies were nominated. As a result, these enemies became abstracted, numberless and therefore unreal (Price 2010: 187). The virtual enemies generated by the War on Terror (like the literally virtual enemies of *America's Army*) serve a similar function to the virtual alliances synthesized by social networking sites and reality television; but their attempts to compensate for a democratic and civic deficit effectively exacerbate the political and social crisis. The enmities and friendships imagined to reinforce the ruins of democracy can only sustain an illusion of that democracy, an illusion which may prevent the possibility of its reinvigoration.

An ideal concept of friendship may be necessary for the possibility of the political and its practice may be symptomatic of the democratic, but its simulation reveals a utility which undermines the essential and definitive afunctionality of friendship in its absolute sense: friendship must be for its own sake; if it is pursued for political reasons it is no longer friendship, and can therefore no longer serve these purposes.

The Revelation of Secrecy

Among its various glosses thereof, Samuel Johnson's *Dictionary* (1755) defines an *intimate* as 'one who is trusted with our thoughts'. Michel de Montaigne (1991: 215) also argues that within the intimacy of absolute friendship there can be no indiscretion: 'That secret which I have sworn to reveal to no other, I can reveal without perjury to him who is not another: he *is* me.'

Yet when we publicly tag our Facebook friends in compromising photographs – when we simultaneously announce and denounce our intimacy – we seem further from Montaigne's idealism than from Friedrich Nietzsche's typical pragmatic cynicism on the subject: 'There will be but few people who, when at a loss for topics of conversation, will not reveal the more secret affairs of their friends' (Nietzsche 2008: 171).

Derrida (2005: 62) points out that, for Nietzsche, friendship displays the semblance of intimacy but lacks 'actual and genuine intimacy'. Friendship is therefore a matter for Nietzsche not of intimacy but of intimation and innuendo. Aristotle and Montaigne create a realm of silence around friendship (nothing is intimated beyond the limits of its intimacy) which allows an uncompromised and unassailable mode of communication within the friendship; Nietzsche, by contrast, permits external intimation (which is the betrayal of intimacy), but cannot countenance such openness *within* the intimacy.

Nietzschean friendship, like Facebook friendship, is therefore a matter of utility which depends upon the intimation of a relationship and upon the simulation of selfhood. 'Is it in honour of your friend that you show yourself to him as you are?' asks Nietzsche (1969: 83). 'He wishes you to the Devil for it!' Friendship is for Nietzsche sustained not by openness and intimacy *per se* but by maintaining the pretence of these things. This mode of friendship is not, however, sustainable in itself, because it is unequal, self-serving, arbitrary and indiscriminate. This mode of friendship recalls one censured by Samuel Johnson (1971: 584):

> His friendship was [...] of little value; it was always dangerous to trust him, because he considered himself as discharged by the first quarrel from all ties of honour or gratitude; and would betray those secrets which, in the warmth of confidence, had been imparted to him.

This Nietzschean notion of friendship as a public and treacherous intimation of trust may not be entirely unfamiliar to the frequent user of Facebook (to those for example who tag their friends in embarrassing photographs) but seems antithetical to the ideal prescribed by Aristotle, Cicero, Montaigne and Derrida, a mode of friendship described by Maurice Blanchot (1997: 291) as one 'which does not allow us to speak of our friends, but only speak to them.'

The Loss of Virtue

In October 2009 the Auschwitz memorial launched its own Facebook page. Auschwitz's Facebook page offers the reader advice on its use: 'Please respect the memory of this place as well as the memory of the victims and the survivors of the camp.' Yet can this simulacrum of human interaction really hope to maintain the integrity of the memory of the victims of the Holocaust?

The BBC reported in October 2009 that Facebook had launched a feature whereby friends and family of deceased users of the site the option to 'memorialize' the profiles of their loved ones. As the BBC added, this move had followed 'some cases of members receiving updates about dead friends.' Facebook's Head of Security Max Kelly wrote in a blog entry of 26 October 2009:

How do you deal with an interaction with someone who is no longer able to log on? When someone leaves us, they don't leave our memories or our social network. To reflect that reality, we created the idea of 'memorialized' profiles as a place where people can save and share their memories of those who've passed.

When one's life is virtual, one's death is simply the inability to log on. The dead are still members of the social networking site: their relationships and memories remain public commodities. Facebook is not letting them go easily: even death cannot disconnect its users from the network. Just as in life the Facebook friendship seems no more than a caricature of the offline ideal of friendship, so in death this hollow imitation of friendship fails to grasp the transcendental mystery (the transformation of friendship into its own ideal) which death might afford.

One is reminded in this context of the journalist and television presenter Victoria Coren's solution to the prospect of the appearance of fake mourners at her father's memorial service: she established and publicized a fictional death – that of one Sir William Ormerod – in order to flush out the coffin-chasing serial funeralists. As Coren explained in *The Observer* in December 2008: 'The death of Sir William Ormerod was announced in July. A tribute website went up, listing Sir William's many achievements in the field of molecular engineering.' Coren's tribute site describes Ormerod as 'an actor and later businessman/philanthropist. He made his fortune developing anti-coagulants for paint, and received a knighthood for services to molecular engineering.' Coren posted announcements of Sir William's memorial service in the classified sections of the broadsheet newspapers and, despite the patent absurdity of Ormerod's achievements, immediately received a string of e-mails from purported friends of the deceased eager to attend his memorial: 'in their greed to get to Sir William's service, these shysters hadn't even waited for the morning newspapers; they emailed in the middle of the night, when the online editions came out.'

The public memorialization of death – made so much more public, undiscerning and immediate by the proliferation of new media technologies – has reinvented, and thereby compromised the authenticity and the integrity of, the process of mourning, a process which the likes of Derrida (1989) and Freud (1984) have seen as the necessarily intimate interiorization of the memory of friendship. That which was supposed to immortalize friendship has ended up killing it off.

There are, of course, undoubtedly cases in which the social networking website may offer a forum for a necessarily (because already) public expression of grief. In, for example, its reports of the murder of a French student in London in July 2008, the BBC revealed that the Facebook page of the victim was 'full of goodbye messages from his friends.' Facebook has, however, also become a site for less well-intentioned memorials to those lost. In August 2011, for example, the *Liverpool Echo* reported that the Facebook page of the convicted murderer of teenager Anthony Walker contained 'the same racist language that was used to taunt Anthony in the moments before his death.' In September 2011 *The Guardian* reported that Facebook user Sean Duffy had been jailed after posting onto the site messages mocking the deaths of teenagers and taunting their families.

In July 2007, during a family holiday in the Algarve, Madeleine McCann went missing just a few days before her fourth birthday. On Christmas Eve 2009 the *Daily Record* reported that:

> A sick Facebook group claiming to be created by Maddie McCann's kidnapper has been removed after thousands complained. The group, named 'If 2,000,000 people join this group, I will give back Maddie McCann', was slammed by outraged users who campaigned to get it banned.

In March 2009 35-year-old Claudia Lawrence also went missing. On 18 February 2010 *The Northern Echo* reported:

> A sick hoaxer who raised false hopes in the long-running investigation into the disappearance of Claudia Lawrence has been arrested. The teenager admitted posting a bogus message on the social networking site Facebook after police swooped yesterday morning. The cruel message appeared on the Claudia Lawrence Investigation page in November last year. It read: 'Hi everyone just let you be aware that I am ok and I am safe and sound. Speak to you all soon. Claudia. xxx.' The message outraged investigators and left Miss Lawrence's already-suffering family in a state of consternation.

We may note in both reports the use of the word *sick*: Facebook hereby becomes the realm not of real friends but of morbid and malicious spectres, impostors and doppelgängers, a virtual parody of the reality and the relationships it purports to reflect and sustain. Aristotle (2004: 200) wrote that 'friendship is a kind of virtue, or implies virtue.' What then when virtue is lost – or if it is revealed that this virtue, this ideal, never in fact existed?

When, in August 2010, a middle-aged British woman called Mary Bale was caught on CCTV dumping a cat into a wheelie bin, the public outcry appeared to concentrate in social networking sites, and seemed rather less focused on feline welfare than upon (often violent) acts of revenge. Facebook erupted with such pages as 'Mary Bale should be locked up for putting Lola the cat in a bin', 'Let's all lock Mary Bale in a bin and leave it in a deserted place', 'Mary Bale Named and Shamed', 'Cats unite against Mary Bale', the 'Mary Bale Hate Group', 'Mary Bale – Stinky and Stale', 'Damn Mary Bale', 'Mary Bale is a cruel bitch', 'Mary Bale – Heartless Old Hag', 'Mary Bale is an Idiot Psycho Bitch', 'Mary Bale is EVIL', 'Death penalty for Mary Bale', 'Death to Mary Bale' and at least twenty similar variants on this theme. As *The Daily Telegraph* reported in August 2010 the 'Death to Mary Bale' page was removed after one Facebook user had declared that the woman should be 'repeatedly head butted' whilst another wrote that 'she should be flogged to within an inch of her life.' Writing of this phenomenon in *The Independent* the following day, columnist Liz Hoggard suggested that while such outrage is perhaps understandable we might engage its energy more productively if we were to 'switch off Facebook and do something different.'

While these instances of outrage and aggression remained in this case at a level of verbal threat and abuse, uses of the social networking site have also led, both directly and indirectly, to other, more physical forms of violence. It was, for example, reported in April 2010 that Londoner Awais Akram had been stabbed and had acid thrown in his face in revenge for an affair he had conducted with a married woman he had met on Facebook.

Facebook advances a public exposure of intimacy which might have devastating consequences upon the relationships involved. The BBC reported in December 2009 that, according to research conducted by Divorce Online, Facebook was being cited in almost a fifth of online divorce petitions. In a 2010 episode of Graham Linehan's situation comedy series *The IT Crowd*, one character, unwilling to tell his wife directly of his plans to divorce her, instead changes his 'Friendface profile from "married" to "it's complicated"' in the hope that 'someone will pass it on.' Earlier, in a 2008 episode of the same series, a husband's observations of his wife's social networking site activities had led him to suspect her infidelity:

> I think Deline might be sleeping with someone else. Little clues here and there. Things only a husband would pick up. Like someone wrote a message on her Friendface wall. Read, 'Can't wait to shag your arse off again soon.' I don't know. Maybe I'm reading too much into it.

Others have responded to such revelations rather more harshly than the husband in Linehan's comedy – and even more harshly than the case of the jilted lover, reported by *The Daily Telegraph* in November 2010, who published sex photographs of his ex-girlfriend on Facebook. In November 2008, for example, it was reported that Wayne Forrester had been jailed for stabbing his wife to death – as a result of her changing her relationship status on Facebook to 'single'. In September 2009 Welshman Brian Lewis was accused of a remarkably similar crime: stabbing his partner to death after she had changed her Facebook relationship status to 'single'. In November 2009 John McFarlane was convicted of murdering a woman with a bolt gun shortly after she had posted a message on Facebook describing his belief that they might have a relationship as delusional. In March 2010 Paul Bristol was found guilty of murdering his former girlfriend after seeing a photograph of her and her new boyfriend posted on Facebook. And these are just a few of the much-publicized examples of such acts – and these few are just British examples.

The case of Raoul Thomas Moat, a former panel-beater and tree surgeon from Newcastle upon Tyne who sparked a massive police manhunt in July 2010, was also conspicuous for its Facebook connections. The news media noted that Moat's shooting spree appeared in part to have been catalyzed by his use of Facebook, and was anticipated by his own Facebook status update: the grandmother of his former girlfriend (and one of his victims) told journalists that Moat had threatened his ex with a gun 'all because she'd put on her Facebook that she was going out with a friend.' The press also noted that immediately before he launched his attacks Moat had updated his Facebook status to read: 'Just got out of jail, I've lost everything, my business, my property and to top it all off my lass has gone off with someone else. Watch and see what happens.' Following Moat's death at his own hands at the climax to a protracted police manhunt, Prime Minister David Cameron condemned public shows of sympathy for the deceased gunman – specifically the messages left for the killer on Facebook. Facebook's memorialization of the dead thus appears as indiscriminate as its modes of interaction with the living.

In his anthropological study of the uses and impacts of Facebook upon the Caribbean island of Trinidad, Daniel Miller (2011: 11) notes that 'when people look at each other's profiles in Trinidad, the thing that they are always alert to is any change in the relationship status that is posted so conspicuously on the profile of every Facebook account.' One might suppose that the pertinence of Miller's observation is not necessarily isolated to Trinidad. Among his case studies, Miller tells the story of a Trinidadian who blames the collapse of his marriage on the pervasiveness of Facebook. Miller (2011: 12) writes that 'Facebook has turned everything into confusion, into public slavering and gossip [...] that sours the relationship itself. It has lost its protection of intimacy and shared secrets.' Again, one might add, this phenomenon is not necessarily exclusive to the Caribbean. The virtualization and publicization of relationships on social networking sites appears to undermine, confuse or corrupt the intimacy of such interactions. The case of Victoria Jones, the teacher who was reported in September 2011 as having stolen baby pictures from Facebook in an attempt to make her ex-boyfriend think they had a baby is a particularly extreme example of this phenomenon. As, for that matter, is the case of another British teacher who was, according to news reports of February 2012, reprimanded by her employers after comments about her social life were seen by pupils on her Facebook site.

The social networking site's confusion of two fundamentally contradictory meanings of the word *friendship* results in the exposure to the public gaze of that which can only be sustained by intimacy and secrecy. This inversion is *uncanny* in a sense which closely parallels that envisaged by Freud (1985: 345): if the uncanny (the *unheimlich*) represents the diametrical opposite of the homely (the *heimlich*), then it is not only the unfamiliar but specifically represents that which should have been kept at home, should have remained private and intimate, but which has been made public. The social networking site becomes an uncanny realm in which secrets revealed rebound to consume the intimacy thus betrayed; a paradoxical realm in which friendship, exposed as pretence, transforms into enmity; an illusory realm whose simulacrum of friendship provokes or masks its opposite.

Facebook effaces the possibility of privacy: Facebook forces all of one's hubris, anger, humiliation and loss into the glare of public scrutiny: it performs, as Papacharissi (2010: 25–50) suggests, a process characteristic of the workings of new media technologies whereby the traditional boundaries between public and private are blurred. The experience of Charlotte Fielder (as reported in June 2011), an amputee who discovered that her Facebook profile picture had been posted onto a pornographic website, is a clear, albeit extreme, case in point – as indeed is the story of Sir John Sawers who, shortly after being appointed head of Britain's secret intelligence service MI6, was publicly embarrassed in July 2009 to find that his wife had posted family details and photographs (with limited privacy settings) onto her Facebook account.

This blurring of the demarcation of public and private space perhaps most overtly undermines conventions of public conduct when it is manifest in relation to issues of national security. As Jackson (2011) has pointed out,

> The UK's Ministry of Defence [...] has launched a campaign to encourage its personnel, and their friends and family, not to share sensitive information. In a video for the campaign, a mother is seen posting information about her son's forward operating base on Facebook – and is then seen having tea with an armed man in a balaclava.

Facebook becomes an uncanny arena for the exposure and the perpetuation of violence and other transgressions. In January 2009, for example, a teenager called Leon Ramsden was convicted of murder, having stabbed a man to death just hours after posting onto Facebook that he felt like killing someone. That December, police were reported as using Facebook status updates in their attempts to track down convicted burglar Craig Lynch, an escaped prisoner who was regularly informing his Facebook friends about the events of his life on the run: from a near car-crash on an icy road to the details of his meals. The following February another burglar, Roy Boodle, was jailed for three and a half years after having used his mobile phone to taunt detectives with Facebook messages during his 18 months on the run from the Greater Manchester Police.

Facebook activity not only provokes, motivates or reveals enmity; it may in fact become a tool in the planning and execution of crimes of extreme intimidation and violence. In February 2009 it was reported that police were hunting one George Appleton, following the discovery of the badly burnt body of his former girlfriend; police noted that Appleton had been known to contact women through such websites as Facebook, Plentyoffish and Informedconsent. In November that year it was reported that Keeley Houghton had become the first person in Britain to receive a custodial sentence for online bullying – having spent six weeks in a young offenders' facility after posting a threatening message about another young woman on Facebook, a message which announced, amongst other things, that she was 'going to murder the bitch.' The following February it was revealed that organized crime boss Colin Gunn had, while in prison, posted on Facebook: 'I will be home one day and I can't wait to look into certain people's eyes and see the fear of me being there.' A week later the then Justice Secretary Jack Straw announced that 30 Facebook pages had been removed because they had been used by convicted prisoners to taunt their victims. The next month, convicted sex offender Peter Chapman admitted to the murder of a 17-year-old girl he had befriended by posing as a teenaged boy on Facebook. Two months later Thomas Mullaney, a 15-year-old boy from Birmingham, hanged himself after being threatened on Facebook.

This tragic and tawdry catalogue is exhausting but hardly exhaustive. The realm of friendship increasingly seems a domain of threat and violence. Such anecdotal evidence also appears to be matched by the statistics. In a discussion of the moral safety of networking websites, Terri Apter (2010: 18) has pointed out that international research has shown that nearly one third of young Internet users aged between 9 and 19 had received unsolicited sexual comments online. If this is friendship, then it is not as we have known it – or at least not as Aristotle, Cicero, Montaigne and Blanchot once idealized it.

On 13 April 2010 the BBC reported that Facebook was 'continuing to resist placing a "panic button" on its pages despite calls to do so by the head of a British child protection agency.' The BBC quoted Richard Allen, Facebook's head of policy in Europe, who commented that the website was one of the 'safest places on the Internet.' On 14 April 2010 *The Independent* newspaper reported that Facebook had been criticized by senior British police officers over its refusal to adopt a ubiquitous 'panic button' – adding that the UK's Child Exploitation and Online Protection Centre had received 253 reports relating to paedophile 'grooming' activities on Facebook during the first quarter of 2010. On 11 July 2010 it was reported that Facebook had finally capitulated to campaigners' demands: that Facebook would introduce allow a 'panic button' application, so that children might report abuse to the UK Child Exploitation and Online Protection Centre. One might assume that Facebook's resistance to the imposition of the 'panic button' mechanism resulted from a concern that the appearance of such an icon would undermine its utopian illusion of a virtual community founded upon an ideal of friendship and openness, an illusion which continues to underpin Facebook's own public image and its defining vision of itself.

The company's similar resistance to calls for more rigorous modes of privacy assurance on the site may similarly relate to its own ideal of itself – a networking platform which excludes malice and criminality and which therefore has no need for privacy or safety concerns. It is unclear whether this apparent naïvety (exposed, for example, by Ron Bowes, the security consultant who in 2010 published the private details of 100 million Facebook users he had harvested from the site) is truly ingenuous. Such cases as that of the Israeli soldier – reported on the BBC on 17 August 2010 – who posted onto Facebook pictures of herself posing with bound and blindfolded Palestinian prisoners, or that of the malicious application (reported by the BBC the previous day) which gained access to Facebook profiles (from which it posted spam messages), seem to suggest that this domain has passed its age of innocence. In August 2010 Google CEO Eric Schmidt told *The Wall Street Journal* that he did not 'believe society understands what happens when everything is available, knowable and recorded by everyone all the time.' The newspaper added that Schmidt predicted that every young person would in the future be automatically entitled to change their name on reaching adulthood in order to distance themselves from the permanent records of 'youthful hijinks stored on their friends' social media sites.' It seems that the demise of that most simple mode of innocence – the innocence of privacy – may lead towards a more profound loss of moral integrity on a societal scale. We can, in short, be exposed to too much information. We need look no further than the case of Kimberley Swann (who in February 2009 lost her office job after having described her work as boring on Facebook) to witness this.

In April 2012, shortly before the closure of an internationally noto-rious website which featured pornographic images of its posters' former sexual partners, the site's owner Hunter Moore told the BBC: 'at the end of the day I feel like I'm just educating people on technology.' Mr Moore's possible ironies or hypocrisies aside, it seems increasingly clear at least that communities of friendship are not the only aims of online social media.

The mass media fixation upon these new-media-related transgressions – big and small – not only testifies to the media's ongoing self-obsession, but may also reveal an anxiety experienced by more traditional media organizations in relation to the applications of emergent media technolo-gies which may threaten to supersede older forms and institutions, a topic which will be explored in more detail in the next chapter.

The Redetermination of Democracy

The Facebook generation seems to found friendship upon a lesser and a less materially committed ideal (a virtual network) than that performed in real-world relationships. This contrasts with, and reverses attempts by, the Aristotelian tradition in social philosophy to base the concept of friendship upon a higher or more ambitious ideal than that generally encountered in the physical world, in order to offer our private relationships as models to which our public lives might aspire. Not only does it set a lesser goal for friendship than the metaphysical ideal: it seems to set its target lower than everyday friendship in a non-virtual society. Rather than aiming towards the philosophical heights, it ends up aiming lower than its actual starting point.

Why then should this matter? Does the commodification of friendship, does the exploitation of the name of friendship, this essential shift in its meaning, the loss of this ancient ideal, really make any difference to our lives at all? Only insofar as (according to some) the traditions of society, politics and democracy are, in western civilization, intimately grounded in this ideal: and to lose sight of this ideal, this aspiration, would therefore be to diminish the possibilities of social, political and democratic existence as it is imagined within a traditional model of European modernity. Friendship represents an inalienable intimacy founded upon mutual transparency, trust, respect, equality and secrecy. It is an immanently private condition which affords a model for public, social and political conduct.

The uniqueness and privacy central to this concept of friendship is uncannily exposed by the blurring of the public and the private provoked, as Papacharissi (2010) argues, by the modes of social interactivity promoted by new media technologies. Papacharissi (2010) suggests that these technologies are redetermining traditional boundaries between the private and public spheres, and redefining social experience as a public expression of one's own private interactions rather than as an activity regulated by the processes and responsibilities of civic participation.

The traditional notion of friendship has allowed the intimacy, transparency and authenticity of the private sphere to be projected as a model and an ideal for the relational conduct of the public sphere. Papacharissi (2010: 133) has proposed that 'the privacy of one's being feeds individual authenticity and self-actualization in ways [...] important to enacting behaviours in public.' We might therefore postulate that the increasing publicization (and therefore the destabilization) of formerly private and intimate modes of interaction may threaten to undermine historically entrenched traditions of public, civic and political relations – relations which have defined such structures as those which underpin democratic pluralism.

So much, then, for the promises of pluralism and dialogue advanced by these interactive technologies. New media have promised to promote the expression and interaction of individuals, and yet they also seem to assimilate and transform difference into a seamless uniformity. The social networking site's cult of the individual is one which standardizes each individual into just another virtual narcissist.

Facebook, electronic government, *America's Army* and *Big Brother* appear at once to be perpetuating western ideologies, structures and perspectives on a global scale – and yet, at the same time they radically undermine the logic of these paradigms (or the democratic values which have sustained them). The times, they are relentlessly telling us, they are a-changin' – but is this a process of liberation or an entrenchment of economic exploitation and socio-political control in the guise of liberation? And might such liberation constitute a process of advanced or revitalized democratization (and, if so, why do we always return to democratization as a moral or teleological absolute, the yardstick of liberation? – does anyone have a better idea?), or commercial hedonism or cynicism, or anarchic individualism, nihilism or narcissism: a simultaneous process of public fragmentation and private homogenization, heralding a state of uniformity without unity (which hardly, from here, sounds like a brilliantly better idea)? Is this shift truly egalitarian or might it merely represent a rebalancing of hierarchy in favour of an ephemeral and depthless populism?

And so we return to this volume's perennial question: are we empowered by these new technologies, or do we just feel more powerful? Do we *know* more, or do we just *think* we do?

CHAPTER 6

Public Knowledge

Another (not entirely unrelated) question: is there any authenticity without authority? Is a structure of authority necessary for the recognition and validation of authenticity?

If (as Jean Baudrillard might suppose) history now appears to have become a depthless, insubstantial image of itself, what then is the status of that commodity which we call *knowledge*? Where does knowledge reside – and can we still learn from it?

Wikipedia

Martin Hand (2008: 15) has, amongst others, suggested that the digital age has been greeted with great optimism by its enthusiasts and denounced with equal passion by its detractors:

> Most commonly, narratives of digital culture imbricate western models of democratization with enthusiastic accounts of information technologies. For some, such technology is instrumental in broader restructurings of modern society, replacing structure with flow, state with network, hierarchical knowledge with horizontal information [...] For others, the use of the term [...] 'digital culture' is hasty or simple determinism, reifying either information or technology as great levellers, an ideological rhetoric which has the effect of glossing an increased penetration and 'hardening' of global capitalism.

Digital democracy does not, from this perspective, appear to have increased civic or political participation or democratic accountability (although it may give a jaded electorate a sense of such participation and accountability). Digital and interactive modes of popular entertainment – from the video game to reality television – have not substantially enhanced the agency of the user-consumer (though they may give the impression that they do). Social networking websites have not as yet ushered in the brave new age of a barrierless society based upon a dialogical structuration of social responsibility. What then of the area upon which the revolutionary potential of information technologies might be expected to be most closely focused: information itself? Have these technologies deconstructed the hierarchies traditionally associated with access to – and generation of – information?

In this context, it seems inevitable that one examines the Internet's – and indeed the world's – foremost source of knowledge, that global phenomenon known as Wikipedia: 'the largest and most popular encyclopedia in the world' (Anderson 2011: 12). It is difficult to imagine a more pervasive information source: Broughton (2008: xv) has reminded us that editions of Wikipedia exist in more than 250 languages – with around four million articles and 20–40 million words added each month in the English language edition alone (Lange et al. 2010: 5; Ayers et al. 2008: 4). As Stewart (2011: 144) has pointed out, almost one tenth of the entire world's Internet users use Wikipedia. As Wikipedia itself announced in 2012, it boasts some 365 million readers worldwide and each month enjoys nearly three billion page hits from the United States alone.

Launched on 15 January 2001, Wikipedia has revolutionized not only our access to knowledge but also our notion of what knowledge is. Despite its avowed requirement to ground all of its content upon authoritative sources – couched in its own imperative that 'all material in Wikipedia articles must be attributable to a reliable published source' – Wikipedia has become notorious for its inaccuracy, while its popularity has simultaneously boomed. This apparent paradox parallels the typical academic response to the site – a combination of public disapprobation and extensive private use (sometimes, one fears, to the point of dependency). It may not simply be the case that, while one recognizes the platform's unreliability, one enjoys its convenience; the ubiquity of Wikipedia has made its unreliability almost an irrelevance. Insofar as the recognition of discourse as knowledge is an act of consensus (or a process of passive acceptance mediated by structures of power), the content of Wikipedia has become 'knowledge' precisely and only by virtue of its presence on that platform. At the same time, however, our acknowledgement of the content of Wikipedia as knowledge is unprecedentedly provisional: the authoritative status of the site remains ambivalent, and users of Wikipedia appear to be more aware than, say, visitors to the *BBC News* website of the unreliability of the knowledge that this platform supplies them, while also being paradoxically willing to accept Wikipedia as their primary (or, for some, only) source of such knowledge. As such, Wikipedia is posited as a definitive authority which lacks authority; and the fact that for the majority of its users this does not appear to be an insurmountable problem suggests that Wikipedia may be promoting (or may at least be symptomatic of) an increasing mistrust of epistemic authority.

In a blog entry of February 2006 the renowned cybersceptic Andrew Keen wrote that 'the truth about Wikipedia, the unintended consequences of its radical democratization of knowledge, is that it turns everyone into kids.' It remains unclear whether, as some might suggest, this destabilization of received knowledge, the consequent disruption to the authority of professional expertise and the growth of what Keen (2008) famously described as the cult of the amateur herald the beginning of the end of polite civilization as we know it – or whether, as the followers of that epistemological historian Michael Foucault might suppose, this phenomenon might prompt a dynamic restructuration of the way in which discourse mediates knowledge and power, and, in doing so, promote a perspective which recognizes the inadequacy and absurdity of that which we call 'knowledge' and which therefore subverts the authority of knowledge and of all authority which is based upon knowledge (which is all authority). The former perspective predicts a future of cynical distrust; the latter anticipates a state of sceptical mistrust. This is perhaps the difference between anarchy or demagogy or autocracy on the one hand (depending on what might result from the feared demise of representational politics), and democracy on the other.

Robert Cummings (2009: 1) has thus distinguished between those perspectives which denounce 'the evils of Wikipedia with the inevitable accompanying lament on the fall of standards and credibility associated with the waning of print culture' and the notion of Wikipedia as 'a harbinger of a new way of writing – and of working.' While some might not share Cummings's apparent enthusiasm for this new mode of cultural production and reproduction, it is difficult to suggest other than that Wikipedia and its ilk are here to stay, and that their influence upon culture and society will continue to be ubiquitous and intense – and that, as we can neither deny nor reverse their impact, it is urgent that we engage with these forms, on their own terms, to understand the nature of that impact. This chapter therefore specifically explores Wikipedia in its own terms: it examines how Wikipedia defines and determines itself, and thereby investigates the modes of culture and society which Wikipedia might represent and propagate: a system which is, as O'Sullivan (2009: 1) suggests, 'alien to our cultural traditions [...] operated on the whole by ordinary people rather than academics or professional writers, whereas we live in a society in which the authority to pronounce on matters of fact of any complexity is regarded as the province of experts.'

John Broughton (2008: xv) notes that 'Wikipedia has never lacked skeptics. Why expect quality articles if everyone – the university professor and the 12-year-old middle school student – has equal editing rights? [...] How can tens of thousands of people work together when there is no hierarchy to provide direction and resolve disputes?' Broughton (2008: xvi) however responds, in defence of the site, that 'Wikipedia has a large number of rules about its process [...] there are a few editors with special authority to enforce the rules.' While Broughton argues that Wikipedia functions by consensus, Wikipedia is then, in many respects, an elitist hierarchy not so different from those in the traditional lifeworld: it does not break away from this problematic model so much as it merely adds other problems, problems of authenticity and formal accountability, to that model.

Joseph Reagle (2010: 152) has compared the revolutionary potential of Denis Diderot's *Encyclopaedia* to that of Wikipedia: 'Much as the *Encyclopédie* challenged the authority of the church and state and recognized the merit of the ordinary artisan [...] Wikipedia is said to favour mediocrity over expertise.' Dan O'Sullivan (2009: 23) has pointed out that Wikipedia resembles the Library of Alexandria as a project aspiring to accumulate all of the knowledge in the world. Does Wikipedia therefore represent the 'new Alexandria' imagined by the likes of Koskinen (2007: 117) and Tiffin and Rajasingham (2003: 26) – or has the information age merely reinforced the relationship between knowledge and power within an entrenched knowledge economy?

Power to the People?

A *wiki* is, according to Wikipedia, a website that 'allows the easy creation and editing of any number of interlinked web pages.' Wikipedia has gone on to point out that there are two kinds of wiki site: open purpose wikis which 'accept content without firm rules as to how the content should be organized' and wikis which 'exist to serve a specific purpose.' In the latter case, Wikipedia has noted, 'users use their editorial rights to remove material that is considered "off topic".' Wikipedia has added that it is itself an example of the latter mode of wiki site, one which works through the evolutionary establishment of a consensus of knowledge, one whose contributors and editors are told not to 'worry about making mistakes' on the grounds that 'no damage is irreparable.' Indeed, Wikipedia has reassured us that 'on average, only a few minutes lie between a blatantly bad or harmful edit, and some editor noticing and acting on it.' It is clearly difficult to see what harm could be done in only a few minutes by a website which is the 'largest and most popular general reference work on the Internet' and which offers 21 million articles, produced by 100,000 active contributors, on subjects ranging from advice on cooking custard through the varying ages of sexual consent in different countries and the mechanics of the jet engine to brain surgery.

Although Wikipedia announces that it 'is not censored' it asserts that 'obviously inappropriate content [...] is usually removed quickly.' The freedom from censorship which Wikipedia promotes is therefore one which may appear to be limited by judgments as to the appropriateness of content to topic (and to other ethical and aesthetic concerns: 'vandalism', is for example, proscribed). It may also be noted that 'content that is judged to violate Wikipedia's biographies of living persons policy [...] will also be removed.' Wikipedia's policy on 'Biographies of Living Persons' demands neutrality and verifiability and requires that such biographical material 'must be written conservatively.' It is, however, somewhat unclear whether (apart from their potential for litigation) there are any strong epistemological reasons why the compilation of information about the living should be more rigorous in its pursuit of accuracy than that about the dead. There are a number of areas within the structure and processes of Wikipedia which some might find similarly epistemologically inconsistent.

There are two mutually exclusive schools of thought on the subject of the contribution to the sum total of human knowledge of the free online encyclopaedia whose content is entirely generated by its dilettante users. One – which we might classify as conservative (one which seeks to maintain entrenched processes for the institutional canonization of knowledge) – would suggest that knowledge is best left in the hands of the professionals: that Wikipedia represents a dilution and a corruption of properly evidenced and provenanced knowledge; for, as Wikipedia has put it in its own description of itself, 'Wikipedia's departure from the expert-driven style of the encyclopedia building mode and the large presence of unacademic content have been noted several times.' Those who hold this position tend to like to cite the many cases of Wikipedia's disastrous errors and embarrassments: many of these (including the blaming of a retired American newspaper editor for both Kennedy assassinations and the premature announcement of the death of a highly respected French scientist), along with the avid media attention they prompted, are well chronicled by Andrew Dalby (2009) in his study of Wikipedia. Some of Wikipedia's more publicly embarrassing incidents include the 2007 case of Ryan Jordan, the Wikipedia editor who passed himself off as a professor of religion at a private university but was in fact a college student who numbered *Catholicism for Dummies* among his primary sources of theological knowledge. That same year Larry Sanger, one of Wikipedia's founding editors, raised questions as to the site's integrity, and went on to launch a new free-access online encyclopaedia (one compiled by experts), Citizendium, in direct competition to Wikipedia. In April 2007 *The Guardian* newspaper's report of Sanger's concerns noted that 'universities have long questioned the reliability of information posted and edited on Wikipedia.' This is perhaps inevitable. One might suppose that such institutions as those unnamed 'universities' which eschew the use of Wikipedia may have vested interests in undermining the credibility of any organization which competes with them in the dissemination of knowledge and information – and whose very existence calls into question the authority of their institutional professionalism.

As Andrew Dalby (2009) points out, the most serious error made by the embarrassed college student Ryan Jordan – in a world in which everyone from professors to schoolchildren are deemed to have an equality of voice – was not to be an amateur but to pretend to be otherwise: 'to make his pseudonymous identity more weighty and academic than his real persona' (Dalby 2009: 138). It is when Wikipedians – and Wikipedia itself – fall back upon an elitism, the institutionally determined hierarchy of knowledge authority which they purport to disavow, that the authenticity and idealism of the project are critically undermined – and its purportedly revolutionary potential is reversed to the extent that it reinforces the entrenched relationship between that-which-is-perceived-as-knowledge and elitist, institutional power.

Wikipedia proudly announces that all Wikipedia aricles 'must be written from a neutral point of view' and yet it fails to observe that the notion of the authoritative primacy of such characteristics as neutrality or objectivity is the primary rationale for entrenched institutional authority: the academic scientificism which denounces subjectivity, and which therefore maintains the myth of the possible existence of objective truth (as if there could ever be a truth independent of subjective perception and interpretation).

In December 2005 the renowned scientific journal *Nature* famously published a paper which compared the website of *Encyclopaedia Britannica* with that of Wikipedia, using 42 'expert reviewers' to carry out this comparison. *Nature* discovered that its experts had discovered errors at a rate of about three per *Britannica* article and about four per Wikipedia article. When *Britannica* queried *Nature*'s findings, the journal responded that its 'comparison was unbiased' and that the journal would 'stand by the story' (*Nature* 2006: 582). The approach taken by *Nature* might be seen as going some way to undermining the institutional or conservative position on Wikipedia (Wikipedia is only a third more inaccurate than *Britannica*); yet, insofar as *Nature*'s study depended upon 'expert reviewers' for its comparison, its perspective remained that of the established and accredited professional: its strategy did not call into question the assumption that what we perceive as knowledge might not require the authority of institutional provenance; nor did it call into question the very notion of accuracy itself. Indeed, in that it was questioning the accuracy of a popular reference work (*Britannica*) rather than that of a peer-reviewed or institutionally recognized academic journal, one might suppose that the primary function of *Nature*'s article was not to laud Wikipedia but to discredit *Britannica* in a manner which sought to reinforce the significance of its own academic authority.

The opposing school of thought on Wikipedia's contribution to knowledge is one which might be described as deconstructionist, and one which draws specifically upon Michel Foucault's influential notion that the categorization of discourse as truth is dependent upon that discourse's relations with the dominant power structures of the culture and society from which it originates and in which it functions: 'it is in discourse that knowledge and power are joined together' (Foucault 1998: 100). The narrative which is acknowledged as knowledge is the one which best suits the hegemonic perspective. A genealogy of knowledge would demonstrate that there has never been any absolute scientific, philosophical or religious truth, but that at any given moment in history pieces of narrative discourse are acclaimed and validated – by academic, cultural, social, political or religious power – as absolute and incontrovertible truths, only to be discredited and re-evaluated as power relations shift into new models. This is necessarily a dynamic process, for, as Foucault (1998: 99) argues, 'relations of power-knowledge are not static forms of distribution, they are matrices of transformations.' From this perspective Wikipedia offers to redefine those areas of discourse recognized as truth with the authority not of fusty institutions of power but of the will of the people. This transformational process might seem at first sight to be terribly democratic.

Yet, as Bourdieu (1977: 184) reminds us, society has moved apparently irreversibly away from a structure in which power relations are determined by subjective individuals rather than by objectified institutions: away, that is, from 'social universes in which relations of domination are made, unmade, and remade in and by the interactions between persons' into a realm of 'social formations [...] mediated by objective, institutionalized mechanisms.' Thus, Bourdieu continues, 'relations of domination have the opacity and permanence of things and escape the grasp of individual consciousness and power.' These formations tend 'to reproduce the objective structures of which they are the product' (Bourdieu 1977: 72). The complicity which such a structuration requires, although sustained by an illusion of individual agency, does not necessitate, support or allow such agency, and may not in itself represent an entirely conscious complicity: its agents (agents as puppets or drones, agents without autonomous agency) may remain unaware of their own role and of their lack of autonomy – the agent's 'actions and works are the product of a modus operandi of which he is not the producer and has no conscious mastery' (Bourdieu 1977: 79). Wikipedia, with its repeated emphases upon its own principles, processes, structures and hierarchies, does not appear to reverse this trend.

For a democracy to function in alignment with (or if it is to develop in the direction of) its defining ideals, individuals must act collectively, but they must also exist and self-determine individually (which they might only come to do through the recognition of their lack of self-determination). The workings of a social networking site may undermine the possibility of democracy insofar as they simultaneously negate the ideal upon which that notion was founded and also fail to recognize the virtual impossibility of democracy (i.e. fail to see democracy as an almost unattainable goal to be pursued through struggle, rather than a right gifted by media institutions). A similar condition of entrenchment may prevail in relation to intellectual as to social capital, insofar as both function as modes of symbolic capital which, as they purport to liberate, effectively bind their subjects to their structures of domination.

We may note in this context that, if Wikipedia were anything like a democracy, then it would be a self-consciously hierarchical one: one whose governance is overtly imposed from the top down. Wikipedia's own page on Wikipedia has explained its structure: 'the Wikipedia community has established a bureaucracy of sorts, including a clear power structure that gives volunteer administrators the authority to exercise editorial control.' It has added that 'editors in good standing in the community can run for one of many levels of volunteer stewardship; this begins with "administrator", a group of privileged users who have the ability to delete pages, lock articles from being changed in case of vandalism or editorial disputes, and block users from editing.' Above the administrators there are 'bureaucrats' (who enjoy the 'ability to add or remove admin rights'), the 'Arbitration Committee' (described as 'kind of like Wikipedia's supreme court'), the 'stewards' ('the top echelon of technical permissions') and Wikipedia's surviving co-founder Jimmy Wales – or as he is now described on the site, its founder – who has 'several special roles and privileges', although the site did not, at the time of viewing, specify what these might be. Andrew Dalby (2009: 10), however, observes that Wales 'holds dictatorial powers to ban users, delete pages and erase page histories.' Dalby also notes in passing that Wales has repeatedly edited from his own biography page on Wikipedia any references to Larry Sanger as the co-founder of the site (Dalby 2009: 148).

The Wikipedia entry on autocracy notes that an autocrat 'is a person (such as a monarch) ruling with unlimited authority' and notes that 'the autocrat needs some kind of power structure to rule.' It is not clear what limits, if any, there are to Mr Wales's power and privileges within his electronic domain, but it is apparent that his position is supported by a rigid structure of power. Indeed, interviewed in *The Independent* newspaper (Burrell 2010), Wales – a great fan of the UK's House of Lords – has compared the site to the British Parliament's upper house: 'He talks of the lack of a written constitution, refers to the website's highest body (its arbitration committee), and notes that "if you become an administrator in Wikipedia, you are pretty much in for life."' Wales has added: 'I'm working as hard as I can to make it as ceremonial as possible ... a constitutional monarch[y] where I have certain reserved powers.'

If Wikipedia has developed a pyramidal hierarchy, then Axel Bruns (2008: 141) observes Jimmy Wales 'at the very top of this pyramid [...] described by some Wikipedians as the site's "god-king".' Bruns (2008: 141) has expressed concerns in relation to the website's increasingly entrenched hierarchization: he argues that Wikipedia's principles of open participation and 'heterarchical governance' may be threatened if, as he puts it, 'current trends towards the development of more permanent administrative structures continue.' Bruns (2008: 143) has added that the incremental development of processes and structures to maintain what he calls the 'heterarchical adhocracies of produsage' may change the nature of the Wikipedia project and thereby undermine the principles of produser (user-as-producer) power upon which Wikipedia is self-consciously founded. It seems that the core pluralist principle of Web 2.0 may be unsustainable: that heterarchy is reformulated as hierarchy to avoid its otherwise inevitable collapse into anarchy.

Joseph Reagle (2010: 119) has, however, defended Wales's position: 'While a founding leadership role has some semblance of authoritarianism to it [...] it is entirely contingent: a dissatisfied community [...] can always leave.' Reagle's argument, similar to that already advanced by Google CEO Eric Schmidt – that unhappy users can vote with their virtual feet – hardly takes into account the virtual monopolies (and therefore the massive agglomerations of power) these organizations hold. Without wishing to draw parallels with some particularly unsavoury extremes of political discourse ('if you don't like it here, you can always go home'), one might also note that Reagle fails to recognize the influence of the socio-cultural *habitus* (as Bourdieu would have it) upon the individual who inhabits and perpetuates it; nor indeed does he acknowledge the power of the *status quo*, a power enforced by the entrenchment of possession and inertia of position. Reagle's argument, were it true, should surely also be true for any political situation, online or off. (Tell that then to the citizens of oppressive regimes.) There is – as Christian Fuchs (2008: 320) suggests – 'a permanent, dynamic self-organization process in which Wikipedia structures and Wikipedians' actions mutually produce each other' – just as, in Bourdieu's terms (1977: 72), the agents of the habitus tend to 'reproduce the objective structures of which they are the product.' This is not then a matter of individual self-determination (although these structures offer the illusion of that) but of the perpetuation of objective systems.

It is fortunate then that Wales maintains his status as – in the words of Reagle (2010: 177) and Bruns (2008: 147) – a 'benevolent dictator' (but then how many dictators were openly described by their subjects as malevolent while still in power?). Wales's position is grounded upon a series of 'policies and guidelines' which support the site's 'five pillars' (its statutes of principle) and which determine the site's content and development. There is undeniably a formal (and relatively fixed) power structure here, albeit one which actively invites participation. One might therefore suppose that, if its development does in fact represent a cultural power shift, then Wikipedia does not convey intellectual authority into the domain of 'the people' so much as it establishes an alternative epistemological elite. As it announces itself, 'Wikipedia is not an experiment in democracy or any other political system. Its primary but not exclusive method of determining consensus is through editing and discussion, *not* voting.'

Wikipedia's own definition of democracy has noted that 'equality and freedom have been identified as important characteristics of democracy since ancient times. These principles are reflected in all citizens being equal before the law and having equal access to power.' Wikipedia may be *free* in the sense that it is free to access; but its hierarchy of roles and rights and its prescriptive principles, however evolutionary, are not designed to promote freedom and equality within the website's own structures. The *Wikipedian* – the contributor to the site – enjoys the illusion of an active participation in the creation of knowledge; and yet, as the site has itself emphasized, 'Wikipedia is not a publisher of original thought' and 'does not publish original research.' Within these limitations, the Wikipedian acts only as an aggregator and disseminator of received knowledge, a gofer to external authorities and internal interests – or, as Samuel Johnson once described the lexicographer, a 'harmless drudge'.

Wikipedia has proudly announced that 'when *Time* magazine recognized You as its Person of the Year for 2006, acknowledging the accelerating success of online collaboration and interaction by millions of users around the world, it cited Wikipedia as one of several examples of Web 2.0 services, along with YouTube, MySpace, and Facebook.' The fact that *Time* magazine chose 'the millions of anonymous contributors of user-generated content' as its Person of the Year emphasizes and reinforces the illusion of significance, agency and empowerment afforded to their users by the likes of Wikipedia and Facebook – an illusion which may (of course) sublimate the desire for real-world participation and allow those users to continue to function as indistinguishable cogs in the post-industrial economy's knowledge machine. Wikipedia has itself explained within what it calls its 'five pillars' – its fundamental principles – that 'no editor owns any article; all of your contributions can and will be mercilessly edited and redistributed.' Wikipedia alienates its workforce from their labour and appropriates the products thereof without even any of the usual material compensations which industrial capitalism once offered. And it expects to be applauded for it.

Wikipedia has described itself as a 'free, web-based, collaborative, multilingual encyclopedia' and it seems significant that it prioritizes the fact that it is *free*. Knowledge may, as Bacon suggested, be power; but it is specifically the ownership of knowledge – the authority of the source of knowledge, the control of the flow of knowledge – which determines the possession of power. Those who own knowledge (or intellectual capital) and control its means of production, mediation and dissemination are, in these terms, rather more powerful than those who merely consume (or for that matter reproduce) it. To be a knowledge-consumer is not a form of empowerment in itself. Wikipedia can supply mediated knowledge for free because that is not the same as giving away power; the power resides in being the knowledge-provider, in being part of (and specifically being at the top of) that organization's hierarchy.

Like those various ongoing initiatives in online democracy and government which have been witnessed across the world, Wikipedia favours and empowers those who have the economic, educational and technological advantages required to access and exploit the possibilities of the Internet; it also promotes the interests and perspectives of those that have the leisure time and the intellectual capital to fill its pages and rise in its ranks. Wikipedia propagates the world-view of a structured and regulated elite, an arbitrary quasi-meritocracy no more open or progressive than that already known in the realms of academic, political and economic capital, and offers this elite the semblance of an extraordinary degree of influence, authority and power.

The Destabilization of Knowledge

Wikipedia's continuing refusal to follow the other global giants of Web 2.0 in their pursuit of financial gain and in their development of commercial interests (Wikipedia refuses to take advertising: advertising, is says, 'doesn't belong here') suggests that the site has not perhaps entirely abandoned the radical potential of its founding ideals. If Wikipedia does not offer an alternative to the relationship between knowledge and power (and, from Foucault's perspective, of course it cannot), then it may yet catalyze a fundamental shift within the very paradigm of knowledge itself: it (apparently unintentionally) may bring to the forefront of our understanding the instability of knowledge which epistemology has always recognized but which knowledge itself cannot usually acknowledge. Socrates is reported to have proposed that s/he who knows s/he knows nothing is wiser than s/he who thinks s/he knows something but actually knows nothing; and Confucius argued that to admit that one does not know something is the heart of knowledge. Could Wikipedia's destabilization of the concept of knowledge therefore in fact stimulate the development of a more sophisticated (that is, a more problematized) popular understanding of knowledge, of its limitations and its uses, which might then advance the basis for the evolution of a more democratic perspective upon authority *per se*? Democracy is based upon trust, but that trust cannot be blind; democracy requires that authority be challenged (and therefore be mistrusted) so that it might eventually and provisionally be accepted (and therefore not be distrusted): in order to avoid a collapse into cynicism, democracy requires the maintenance of scepticism.

It should be noted in this context that Wikipedia is (of course) an ever-changing organism, and, as such, challenges from the outset the academic's desire to cite the hard evidence of permanently fixed texts. Knowledge, we like to think, should be as solid as the bricks and mortar which house its ancient institutions; the revelation that knowledge is fluid may be somewhat distressing to those who have founded their entire careers upon such ambiguous materials. Yet this revelation is key to any notion of widening participation in the processes of knowledge, and therefore in the processes of power. However, although Wikipedia's processes and its inherent and self-conscious mutability may offer to subvert any fixed, formal and deferential notion of the authority of established knowledge, its avowed mission – that of the public encyclopaedia – necessarily relies upon and perpetuates such a sense of authority. In its attempts to be encyclopaedic – to be comprehensive, objective and neutral – it cites established knowledge and its own gate-keeping, self-regulating hierarchies (not dissimilar to those of academia) as the provenance of its authority; but it is through its failure to become a source of knowledge as formally and immutably authoritative as those it cites (simultaneous with its ability to become a source more influential than those traditional sources) that Wikipedia prompts a radical shift in popular perceptions of knowledge and its authority.

When this online encyclopaedia speaks of the 'disambiguation' of its terms (as it so often does) it in effect draws attention to ambiguity, to the polysemy of knowledge: when Wikipedia *disambiguates* it demonstrates the multiplicity of possible meanings and associations applicable to a term; it effectively re-ambiguates. 'Wikipedia is not a democracy' it announces on its own list of things it is not; but there is something at least pluralist – and therefore fundamentally or structurally democratic in a sense of the notion of democracy which stands opposed to demagogy – in its ambiguation of authority.

Yet when we examine Wikipedia's own perspectives on knowledge itself, we return to the intrinsic conservatism which returns constantly to undermine the portal's revolutionary promise. Wikipedia's entry on 'knowledge' has *disambiguated* (or defined) its primary notion of knowledge as 'the possession of information' – distinguishing this concept from knowledge as 'the philosophical concept studied in epistemology', from 'the rigorous geographical training obligatory for London taxi drivers' and from – amongst other things – a Jamaican reggae group, two television channels, a series of children's books and a song by ska-core band Operation Ivy, all of the same name. Knowledge is about possession – that, despite this outpouring of other fascinating facts, is its first and prevalent sense.

The Wikipedia entry on 'knowledge' has opened by quoting the *Oxford English Dictionary*'s definition, one which prioritizes 'expertise, and skills acquired by a person through experience or education.' In the spirit of this authoritative source's authoritative (expert-centred) definition, Wikipedia has then cited Plato's formulation of knowledge as 'justified true belief' – before going on to reference Aristotle, Russell, Rorty and Wittgenstein, amongst others, and supposing that 'the definition of knowledge is a matter of on-going debate among philosophers in the field of epistemology.' The implication is clear: that this subject and this debate is the province of qualified philosophers and epistemologists rather than a matter of concern for the public at large.

As of August 2010, the Wikipedia entry on knowledge included just over 400 words on scientific knowledge, 236 words on religious knowledge, 179 words on situated knowledge (which it described as knowledge based on experience rather than scientific or religious authority) and only 130 words on the partiality – the problematic incompleteness – of knowledge. By May 2012 the page featured 465 words on scientific knowledge, 334 words on religious knowledge, 171 words on situated knowledge and just 108 words on the partiality of knowledge. This loading of emphasis upon the stability, certainty and authority of knowledge (rather than upon its ephemerality) is echoed in the illustrations which accompany the article: a portrait of Sir Francis Bacon (captioned with Bacon's famous maxim that 'knowledge is power'), a Greek statue personifying knowledge in amongst the ruins of the ancient Library of Celsus in Ephesus, and a mural by Robert Reid dating from 1896 and depicting an anthropomorphic idealization of *Knowledge* – a young woman holding a large book. Reid's mural adorns a wall of the Library of Congress in Washington D.C. above a caption which announces that 'knowledge is the wing wherewith we fly to heaven.' The message of Wikipedia's text and its illustrations seems to be that knowledge is for the most part reputable and reliable; it is the product of professional scientists and seers; its discussion is the province of accredited philosophers and academic authorities; its place – indeed, it source – is the library; it is the immutable and absolute key to the divine, to enlightenment and to power (but only via these figures, symbols and institutions of authority).

We may however note (as we may discover from Wikipedia's own entry on the painter Robert Reid) that in the north corridor of the Library of Congress's second floor Reid's representation of *Knowledge* is accompanied by a similarly idealized embodiment of *Wisdom* – beneath which stand the words 'knowledge comes but wisdom lingers.' At the bottom of the Wikipedia page on *knowledge* we find the standard statement on the currency of the entry: 'This page was last modified on ...' Along with the words beneath Reid's *Wisdom*, the derelict state of the antique Library of Celsus, and Socrates and Confucius's maxims on the subject, this statement may serve to remind us of the uncertain and impermanent nature of knowledge – and may remind us that knowledge is as ephemeral as the power which (for Bacon) it promotes and which (for Foucault) promotes it. If Wikipedia is the new Alexandria, then we should not forget what happened there.

Wikipedia seeks to create the conditions for a homogeneous and authoritative consensus of knowledge, a new corpus, canon and world-view constructed by – and initiating – an emergent epistemological elite, a prescribed and regulated hierarchy founded upon, and assuming, the authority of earlier intellectual hegemonies. However, in doing so it exposes the fragility of, and thereby undermines the authority of, the concept of established knowledge – the very myth from which Wikipedia endeavours to develop its own power. Wikipedia is not democratic (as it says it is not); its mission is as authoritative as those structures it seeks to supersede; and yet its failure to achieve its mission opens up possibilities for the promotion of democracy.

Insofar as its contributors lack the trappings and symbols of academic power (the reassuring regalia of dress, title, position, ceremony, architecture, lecture, print publication, media publicity, research funding and professional consultancy) which tend to veil the incoherence of academic knowledge, the absurdity of Wikipedia's operations (conspicuous in their often geekish amateurishness) may be seen as subverting the platform's attempts at epistemic authority. But at the same time Wikipedia has become synonymous with knowledge: it has become the post-industrialized world's first and most influential resource for information. It 'disambiguates' knowledge as 'the possession of information' – knowledge is its possession; and the general recognition of Wikipedia's primacy in the ownership of knowledge is crucial to its authority and power. Wikipedia *is* knowledge: it is the domain of knowledge; knowledge is its domain. Wikipedia is also patently and famously unreliable and ephemeral. Try as it might to establish the authority of knowledge *per se* (in order to establish its own authority, which is based upon that authority) Wikipedia therefore inevitably returns – even in its own attempt to define knowledge – to a position where it calls the authority of knowledge, of all knowledge, into question, and thereby (insofar as knowledge is, for Foucault as for Bacon, inseparable from power) undermines the possibility of the continuation of an unquestioning acceptance of authority itself. As Foucault (1998: 101) has suggested, while 'discourse transmits and produces power; it reinforces it, but it also undermines and exposes it, renders it fragile and makes it possible to thwart it.' This thwarting of the structures of power is not necessarily an advertent process. Despite itself, and despite its hierarchical and even autocratic tendencies, and as an unintended consequence of its own failings and contradictions, Wikipedia may thereby offer an opportunity for the development of democracy, in that the tendency to challenge – and thus to begin to *thwart* – the structures and processes of power is crucial to active and participatory citizenship in a pluralist and democratic society. But it may only be when Wikipedia and its users become aware of this potential (a potential which requires an awareness of the fundamental incoherence – and thereby of the fundamental value – of knowledge and knowledge-based authority) that this phenomenon might manage to defer or prevent the slide into philistinism or into that dictatorship of ignorance which many fear Wikipedia now represents.

Nutopia

Wikipedia, it has informed us, 'was founded as an offshoot of Nupedia, a now-abandoned project to produce a free encyclopedia. Nupedia had an elaborate system of peer review and required highly qualified contributors, but the writing of articles was slow.' Founders Wales and Sanger therefore elected to employ a wiki method of information collection and editing in order to allow for the expeditious provision of extensive free content. There is undoubtedly an idealism at the heart of this project, albeit one which from the start has been tempered by pragmatism in order to develop and sustain the project's momentum.

If Wikipedia has come to represent the sum total of human knowledge, then Wikipedia offers a model or a map of the world which is becoming increasingly equivalent to, which indeed subsumes, the world itself. As such, it has come to resemble the fantasy realm of Tlön, Borges's world imagined at first as a theoretical model by intellectuals which eventually becomes the model on which the material world bases itself (Borges 1970) – or Baudrillard's notion of the map of the world which becomes the world (Baudrillard 1994). In short, the simulacrum has taken over: but this is a simulacrum constructed not by Borges's academics but by a section (admittedly an educationally and economically privileged section) of the general population. What kind of world, then, would such a project construct? Is this new media Utopia to be idealistic or pragmatic, democratic or demagogic, anarchic or autocratic, egalitarian or elitist, meritocratic or vulgarian?

One thing, at least, is clear – Wikipedia self-consciously endeavours to be pluralist: 'we strive for articles that advocate no single point of view. Sometimes this requires representing multiple points of view, presenting each point of view accurately and in context, and not presenting any point of view as "the truth" or "the best view".' In its quest for multivocality Wikipedia offers a mode of practice which attempts to meet an ideal not dissimilar to John Fiske's best-practice model for journalism. Fiske (1987: 307–308) has argued that in a progressive democracy, news should 'nominate *all* its voices' – in other words, it should name all of its sources and therefore never present any of its perspectives as ideologically neutral or natural. Wikipedia has offered a similar strategy:

> Values or opinions must not be written as if they were in Wikipedia's voice. When we want to present an opinion, we do so factually by attributing the opinion in the text to a person, organization, group of persons, or percentage of persons, and state as a fact that they have this opinion, citing a reliable source for the fact that the person, organization, group or percentage of persons holds the particular opinion.

Wikipedia's willingness both to nominate and to invoke a plurality of voices invites the inclusion of those voices which criticize Wikipedia itself. As of summer 2010, while the platform's entry on the topic of 'Wikipedia' accounted for just under 12,500 words, its section on 'Criticism of Wikipedia' (listing externally published criticisms of the site) ran to over 14,500. By spring 2012, however, while the former page had grown to over 17 thousand words, the latter page appeared to have disappeared – a search for 'criticism of Wikipedia' redirected the browser back to the former page.

The site's original (but perhaps diminishing) openness to criticism – and the promotion of a dynamic, pragmatic and self-correcting approach to its own development – would appear to locate Wikipedia within Fredric Jameson's notion of the meta-utopian (Jameson 2005), that mode of utopianism which inscribes its own counter-argument within itself. Indeed, Wikipedia has not only cited external criticisms: its internal plurality of voices also continues to offer its own criticisms of the platform. We might, for example, discover – in the site's section on 'Why Wikipedia is not so great' – the assertion that 'Wikipedia *is* a bureaucracy, full of rules described as "policies" and "guidelines" with a hierarchy aimed at enforcing these.' Does Wikipedia therefore offer the potential to aspire to the pluralist, pragmatic, dynamic, self-deconstructive criteria for the sustainability of what Tom Moylan (1986: 213) has described as a 'self-critical utopian discourse'?

For as long as Wikipedia has been able not only to inscribe and promote criticism of itself, but also to demonstrate its awareness and acceptance of – and draw attention to – its own self-defeating inadequacies and innate contradictions, then it might be seen as having chosen the path of philosophy over that of philistinism, of consensual democracy over demagoguery. This is not, of course, so much an end *per se* as it is an ongoing and dynamic process: one whose participants move continually towards a consciousness of the absurdity and the inevitability of the relationship between knowledge and power, and yet one which may be swiftly undermined by self-censorship and the re-imposition of rigid hierarchies.

As Wikipedia fast becomes the post-industrialized world's primary model of its own knowledge of itself, it seems clear that Wikipedia's politico-societal structural orientation may prove influential in the physical world's understanding and construction of itself. Like all opportunities of any significance, this situation of course poses significant risks. By increasing the speed and spread of the ideological influence of mass media forms, new technologies have exponentially upped the ante. Our chances of a progressive future may depend on our alertness to the risks of the alternatives.

The biggest problem is, of course, that Wikipedia becomes a self-sustaining and self-censoring autocracy based upon an elitist hierarchy of knowledge authority which not only reflects and magnifies the disparities and distortions of academic and informational institutions in the physical world, but which also lacks the historical provenance and public and professional accountability upon which those institutions were founded – and that therefore the information it provides is useless and valueless except as a matter of symbolic exchange which underpins the socio-political positioning of its elite. Andrew Dalby (2009: 177) has pointed out Wikipedia's tendency to offer as its information sources websites which are themselves only mirror sites of Wikipedia (sites which have copied Wikipedia content); in these cases, Wikipedia becomes a depthless, unprovenanced and therefore unquestionable simulacrum of knowledge, a copy or reflection without an original object (and here specifically one without original interpretation or research) – a symptom of the fetishization of knowledge as an economic commodity, rather than an opportunity to democratize not only access to knowledge but also the production, interpretation and recognition of knowledge. Insofar as Wikipedia bases itself upon, and therefore compounds, the structures and principles of traditionally institutionalized knowledge (the authority of an epistemological hierarchy which sustains an illusion of scientific objectivity), it perpetuates and extends the age-old trade in knowledge as socio-economic and political capital – a capital which may lack any real use-value in itself – a symbolic substitute for that value. This is not the democratization of knowledge, but its opposite: the devaluation of knowledge, the replacement in the public sphere of useful knowledge with its simulacrum, a devalued and merely symbolic substitute. It represents the devaluation therefore of the citizen as knowledge-consumer, and also of the citizen as knowledge-producer, in that the individual's own contribution to knowledge (their subjective positioning, their analytical interpretation and their original research) is denounced as invalid, as a corruption of the purity of fetishized information. In these terms, Wikipedia's logo – its incomplete jigsaw globe – might seem to some to recall the half-assembled Death Star in *Return of the Jedi*: a mechanism which, once complete, will devastate all within its path.

The Death of Democracy

Andrew Keen (2008: 7) has argued, in his best-selling polemic against digital media culture, *The Cult of the Amateur*, that 'as traditional mainstream media is replaced by a personalized one, the Internet has become a mirror to ourselves.' The Internet becomes an instrument not for the mediation and dissemination of information, nor for the promotion of social, cultural and political dialogue, but for the narcissistic self-expression and self-publication of its individual users; and one might suggest that this precisely relates to the distance and anonymity of publication: the user can publish herself because it is not really herself, but is rather her online subjectivity, a self without responsibility, commitment, embarrassment or visible audience. When we publish ourselves (or, as YouTube would have it, broadcast ourselves) into the cyberether, we are to some extent merely throwing avatars into an abyss. Keen (2008: 7) has suggested that we broadcast ourselves 'with all the shameless self-admiration of the mythical Narcissus' – and, like Narcissus, we may be consumed by our own reflections. This can hardly be seen as a mode of progressive extrapersonal interactivity.

Though Keen does not deny being an elitist (2008: xvi–xvii), his railing against this narcissistic simulacrum of value and authenticity is not necessarily elitist: indeed he specifically argues against the 'real political reactionaries' in the emergent elite of the digital media industry, the new 'antiestablishment establishment' (Keen 2008: xviii, 13). He berates the Internet's 'souring' of civic discourse and asserts that 'the decline of the quality and reliability of the information we receive [is] distorting, if not outrightly corrupting, our national civic conversation' (Keen 2008: 15, 27). If that is in fact the case, then this would seem like a very undemocratic form of democratization indeed. If populism is rebranded as democratization, then we may come to lose the provenance, depth and meaning of democracy – and political society may thus become no more than a narcissistic simulacrum, an empty sign which stands in place of something which we have valued but can no longer recall. This, then, would be what Keen (2008: 54) dubs 'the degeneration of democracy into the rule of the mob and the rumor mill.'

Indeed, there have been a number of high-profile cases in which the social networking sites of Web 2.0 have become the vehicles for such mob mentalities. The seasoned journalist and self-proclaimed 'hackademic' Ivor Gaber (2010) – in reference to the use of social networking technologies in the campaign against *Daily Mail* journalist Jan Moir's homophobic attack upon the late Stephen Gately (reported, for example, by *The Guardian* in October 2009 as the 'Twitter and Facebook outrage over Jan Moir's Stephen Gately article') – has spoken of the emergence of a demagogic 'Twitter mob rule – a Twitter dictatorship.' Or, as the Conservative MP Louise Mensch told *The Independent* newspaper in May 2012, 'if you want to see the worst of humanity, look on Twitter.' The fact that in May 2012 Twitter topped ten million users in the UK alone may raise some concerns in the context of such observations. (See also Webb 2012.)

These platforms' potential for such abuse was underlined by the news in March 2012 of a custodial sentence being given to Liam Stacey, a student who had posted onto Twitter racially offensive comments about a critically ill footballer. Indeed *The Times* newspaper's crime editor Sean O'Neill has argued that the output of the Twittering classes may pose a threat to the equitable processes of the justice system (as reported by *BBC News* in May 2011): 'It could prejudice a trial, put a protected witness's life in danger or cause serious psychological damage to a victim of sexual assault. All because so-called citizen journalists don't know what they're doing.' One particular case of online social networking proving prejudicial to the process of justice came to light in June 2011 when a former juror who had contacted a defendant via Facebook (thus causing a trial to collapse) was jailed for eight months for contempt of court. The social networking site's blurring of traditional boundaries within the justice system was also witnessed in March 2009 when prison officer Nathan Singh was dismissed from the

service after befriending criminals on Facebook – and in spring 2012 when
a number of people were arrested after posting onto Twitter the name of
the nineteen-year-old victim of a rape by a Welsh footballer. While, in
their defence, it may be argued that such sites also serve as an antidote to
hypocritical super-injunctions which affords the dissemination of inanely
scandalous rumours as to the personal lives of the rich and famous, it may
not be entirely obvious to some how such activities directly promote the
public interest.

British Summer, Arab Spring

Yet in their ability to subvert institutional structures, might not such sites
have a role to play in the specific subversion of oppressive structures of
governance? Some credence has been given to the idea that such sites as
Twitter facilitated and expedited the mobilization of pro-democracy dem-
onstrators during the so-called Arab Spring, the wave of protests against
authoritarian regimes which began in December 2010. It remains unclear,
however, to what extent this emblem of westernization was seized upon
by the western media as a way to make the story more accessible and interesting (and of course to facilitate media access to original source material,
questionable though some of these sources might have been) – and to what
extent these social networking technologies may be seen as promoting and
enhancing civic discourse itself. The Libyan author Hisham Matar has, for
example, argued that the role of social media in the Arab Spring has been
overstated by media commentators (Singh 2011). Matar has suggested that
'the Egyptian uprising didn't happen on Facebook or Twitter because it
couldn't have happened without the working classes, and they don't have
access to those things. But it allowed the agile, internationalist elite to
mobilise and play to the international media.' Matar added: 'Social change
takes a very long time. The Internet is one of very many different tools and
I don't think it's always going to make or break an uprising.'

As Chebib and Sohail (2011: 155) have suggested, 'social media itself cannot be termed as a trigger for the revolutions.' They stress that in the Egyptian uprising of 2011 'social media's main role was as a facilitator and an accelerating agent.' Courtney Radsch (2011: 80–81) points out that in the months leading up to Egypt's uprising the Egyptian blogosphere, while reflecting a political situation that was clearly 'combustible', lacked a revolutionary spark – and that this spark was provided by the Tunisian uprising – despite that fact that such combustibility was not particularly evidenced in Tunisia's own blogosphere. We may infer, therefore, that it was not blogosphere itself which set the region alight. Indeed Morozov (2011b) has added that, while 'it's been extremely entertaining to watch cyberutopians [...] trip over one another in an effort to put another nail in the coffin of cyberrealism', those cyberutopians who think the Arab Spring was ignited by activities on social networking websites are ignoring 'the real-world activism underpinning them.'

In September 2011 BBC journalist Mishal Hussain presented a two-part documentary about the Arab Spring entitled *How Facebook Changed the World*. What is notable about Hussain's documentary is how (despite its title) she demonstrates that social networking sites were perhaps most significant not in fomenting revolution internally but in their capacity 'to show the outside world what was happening.' Hussain also repeatedly emphasizes the limitations of the virtual revolution. She points out that in Egypt only 20 per cent of the population had Internet access, and explains how messages were sent not through the electronic ether but via taxi drivers. As Hussain stresses, when the Egyptian authorities blocked the Internet, 'the activists already had their plan and technology was no part of it.' She adds that when the Libyan government made a similar move, Libyan rebels reverted to 'old technology' – driving to the border with their video footage to get their messages out. Hussain's documentary concludes with a broader question – and one which remains as yet unanswered – the question as to whether the same technology that helped these nations break with the past 'can be harnessed for a better future.' The eventual outcome of these protests – the prospective rise of a radical (and possibly Islamist) demagoguery – may make those western cyberenthusiasts who have deployed the notion of the crucial use of such technologies in the propagation of the Arab Spring to support their arguments in favour of the democratizing powers of Web 2.0 reconsider their positions. Indeed the violent political and religious clashes that have erupted in Egypt since the country's revolution have seemed to underline this point. As Ghannam (2011: 23) has suggested, 'blogging and social networking alone cannot be expected to bring about immediate political change' and that we should therefore focus not upon the headline-grabbing online drama but upon 'the long-term impact, the development of new political and civil society engagement, and individual and institutional competencies.'

At around the same time a number of anti-capitalism protests had begun in the UK, particularly in the area of St Paul's Cathedral in the City of London. Similar protests were seen on Wall Street in New York. Commentators questioned the point of these protests – not their sincerity, but the fact that they did not have a defined goal. Like the protests against economic reforms that had regularly engulfed Athens for the previous few years, they appeared more like an outpouring of anger than a focused attempt to drive a particular political or economic agenda. These clashes and demonstrations – in Cairo, London, New York and Athens – seem to demonstrate that while new media technologies clearly are useful as communication tools in mobilizing protesters to action, their usefulness in establishing strands of dialogue which might coalesce into specific strategies – in other words, their promotion of a public sphere of socio-political debate – seems limited. A similar phenomenon was also demonstrated by Barack Obama's first presidential election campaign's success in its mobilization of support, but its eventual failure in the development and establishment of an agenda-setting consensus.

Peter Jackson has written of the uses of social networking sites in the fomentation of incidents of civil unrest which few would judge as instances of civil liberation: the series of riots which engulfed English towns and cities in August 2011. Jackson (2011) has suggested that 'as riots spread across England [...] so too did a wealth of misinformation about them, fuelled by social networking sites like Twitter [...] As the rioting moved to other parts of London and England, inaccurate – at times inflammatory – information

began appearing on social media sites [...] In times of uncertainty or heightened fears in communities, rumour mills have always churned. But in a
world of social media, a whisper can acquire a damaging momentum regardless of its relationship to the truth.' Indeed, a number of individuals have
been jailed for their use of social networking sites to incite these riots: one,
for example, had launched an event on Facebook entitled 'Smash down
Northwich town' while another had started a Facebook page called 'Let's
have a riot in Latchford'. Yet another was given a four-month custodial
sentence after suggesting 'Let's start Bangor riots' on Facebook. One particular judicial decision led that August to the identification of a further
inciter as 16-year-old Johnny Melfah (courts would not normally name
minors involved in criminal prosecutions): according to the BBC, Melfah
'admitted inciting thefts and criminal damage on Facebook.' Towards the
end of that month the British government announced a summit meeting
between senior police officers and representatives of Facebook, Twitter
and BlackBerry to explore, in the words of a Home Office spokesperson,
'whether and how we should be able to stop people communicating via
these websites and services when we know they are plotting violence, disorder and criminality.'

Summer of Discontent

The events which began in Tunisia in December 2010 and which spread
across the Arab region have begun to foster some measure of a renaissance in
cyberutopianism. Such luminaries as Stuart Allan (2011) and John Downing
(2012) have reiterated an optimism for the democratic potential of new
media in response to the uses of such technologies by the democratic revolutionaries in this region for the mobilization of popular action. Khamis
and Vaughn (2011), for example, argued that 'the success of the Egyptian
revolution, and the effective role that new media played in it, has broad
implications [...] throughout the world.'

The news media have also pushed this agenda: Baker (2011), for example, wrote in *Time* magazine that 'members of the new opposition [...] put up Facebook pages and posted on Twitter [and] exhausted their thumbs sending text messages to everyone in their mobile phone books.' Williams (2011) wrote in the *Daily Mail* that 'as in the uprisings that toppled long-time autocratic rulers in Egypt and Tunisia, Libyan activists are using social networking websites like Facebook and Twitter to rally protest.' Indeed Mark Almond (2011) penned a piece for *The Sun* newspaper unambiguously headlined: 'How social networks trigger political downfall'.

Robert Fisk (2011) announced in *The Independent* that 'this is revolution by Twitter and revolution by Facebook.' Hugh Macleod (2011) in *The Guardian* reported on 'Syria's switched-on cyber activists.' Fawaz Gerges (2011) argued in *The Mirror* that the Arab 'revolt is powered by [...] cellphones, the Internet, Facebook and Twitter.' *The Times*'s Stephen Dalton (2011) suggested that Twitter was 'proving mightier than the sword throughout the Middle East' – while the same newspaper's Adam LeBor (2011) proposed that '140 characters can spark a revolution.' Indeed, on 13 February 2011 an editorial in *The Sunday Times* predicted that 'more revolutions will be fuelled by Twitter.'

Martin Fletcher (2011) in *The Times* reported from Egypt's Tahrir Square, where 'young men spray-painted Google, Twitter and Facebook logos on walls and tanks.' Fletcher referred to computer-technician-turned-revolutionary Wael Ghonim as 'the Facebook dreamer who led his nation.' Indeed Ghonim was later described by Ed O'Loughlin (2012) as a revolutionary who 'felt the need to tear himself away from the action because he wanted to update his Facebook page.' Though Ghonim (2012) would later emphasize the role of new media in his account of what he calls *Revolution 2.0*, his words that day in Tahrir Square were perhaps more illuminating: 'I liked to call this the Facebook Revolution, but after seeing the people out there I think it's the Egyptian people's revolution' (Fletcher 2011).

In his memoir of the Egyptian uprising Ghonim refers repeatedly to the shortcomings of new media in terms of their revolutionary effects. He notes for example that members of online campaigning groups tended to be young while older Egyptians remained away (Ghonim 2012: 113). He points out the problems faced by the online revolutionaries when Facebook chose to suspend certain campaigning pages because of copyright violations or fake user identities (the latter had of course been considered essential to the campaigners' safety), noting, however, that those online conspiracy theories which suggested that Facebook had struck a deal with the Egyptian government to block certain activists' pages eventually proved baseless (Ghonim 2012: 113–119). He emphasizes that, while the posting of personalized content on social networking sites can inspire people in ways that the facts and statistics generally used by human rights campaigners cannot, he is not suggesting that the former could ever replace the latter (Ghonim 2012: 88). He also speaks of the need to use these sites not only for the purposes of 'communication and coordination' but also to 'promote a culture of dialogue [...] and to cultivate a tradition of tolerance' (Ghonim 2012: 155, 113). Ghonim's problematization of these media forms has not, however, entirely diminished the cyberenthusiasm of some western commentators.

The western media's attempts to appropriate these revolutions as offshoots of western media technologies and cultures may to some extent be explained by their desire to atone for (or to cast a veil over) the apparent indifference of many western media organizations (and many western governments) to the decades of human rights abuses perpetrated by these Arab states, or, at worst, to allow the West to forget its complicity in these regimes: from Tony Blair's now notorious public and secret visits to Colonel Gaddafi to the case – revealed by *The Independent* in February 2012 – of the BBC's screening of a documentary made by a company funded by the Mubarak regime. There seems something meanly self-aggrandizing in the West's claims to the liberatory potentials of its own technologies. For, as Doreen Khoury (2011; 84) has supposed, the 'ownership of the Arab revolutions will always belong to the Arab people and not to Facebook or Twitter or any of the other online tools.'

Furthermore, it has become increasingly clear in these post-revolutionary nations that Web 2.0 has not established a dialogical political consensus, and that the public sphere, such as it is, remains a violent, turbulent and resoundingly material space. Within this context, it seems that new media technologies may not have represented a primary cause of democratic agitation (as a site for the development of political consciousness, debate and consensual strategy) so much as a practical catalyst, a useful tool for the mobilization of flash demonstrations and for the dissemination of information and images to international news organizations (and indeed one which could be replaced by more traditional, low-tech tools when the Internet was not available).

New media have also been cited (by certain elements in the news media and in politics) as the cause of the riots which began in London on 6 August 2011 – and which then spread across other British cities (including Liverpool, Manchester, Birmingham and Bristol) over the following four days. During this short period the *Daily Mail* newspaper, for example, out of its 67 stories about the riots, featured 16 stories which held new media in some way accountable for this outbreak of urban violence (including 4 stories which focused on new media as a primary agent in this unrest) – as opposed to only 14 stories which blamed left-wing and liberal social politics (a traditional target for the *Mail*) and only nine stories which blamed the police (and just three which appeared to portray the BBC – another perennial *Mail* target – in some way at fault). In only seven stories during its immediate coverage of the riots did the *Mail* refer to new media in a less than critical fashion.

Analysis of *The Sun* newspaper's coverage of the riots over the same period shows similar results. Only two of the 58 stories which its national edition ran on the subject blamed the political left – while 11 portrayed new media negatively in relation to the riots (and only four gave new media a positive spin). The Scottish edition of *The Sun* was even more forthright in connecting the riots with new media: over this period its headlines included 'Thieves used Twitter to urge looting' (8 August), 'Nail the Twitter rioters' (9 August) and 'Scots pair held over Facebook riot pages' (10 August; see Figure 2).

The emphasis on the role of interpersonal media technologies in these events is somewhat unprecedented. Commentaries on the Brixton riots of the 1980s did not, for example, focus on the telephone as the cause of the unrest, nor has the early postal system been commonly held directly to blame for Peterloo. In part it may be supposed that certain elements in the British press chose to blame the uses of new media technologies for this civil unrest in preference to addressing such issues as a growing sense of socio-economic injustice among a disaffected and excluded youth as possible root causes of these events (a phenomenon, for example, identified by *Reading the Riots*, a study conducted by the London School of

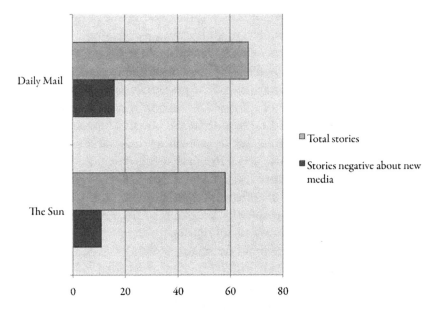

Figure 2. Negative coverage of new media in relation to total coverage of the summer riots in *The Sun* and *Daily Mail*, 8–11 August 2011.

Economics in collaboration with *The Guardian* newspaper). In the ensuing hysteria the popular press (those age-old voices of reconciliation and reason) seemed entirely unaware of their own role in the spread of panic and chaos across the nation.

It may also be supposed that the UK's traditional news institutions' ambivalent but emotive perspective on new media technologies may to some extent explain their praise of the influence of interactive technologies in Arab rebellions and their almost simultaneous condemnation of that influence in British riots. Journalistic organizations, while obliged to get into bed with new media technologies (in their exploitation of online publication, for example, and of user-generated content), appear to continue to feel professionally, institutionally and economically threatened by the rise of this prospective Fifth Estate. While Jones and Salter (2012: 173) may suggest that the future of journalism lies in its integration of online technologies, a certain degree of scepticism (and dread) clearly remains.

A similar ambivalence appears to apply to academic perspectives on new media: perspectives which are at once optimistic about the potential of new media to promote a democratic and egalitarian future envisaged by the liberal ideals dominant in so much of humanities and social science academia, and enthusiastic as to the potential of new media for the dissemination of the fruits of academic endeavour (through the technologies' potential to sponsor a far broader reach of education and publication than previously possible), while at the same time cautious as to new media's potential to de-institutionalize or de-professionalize intellectual and epistemological authority. Just as the personal news-blog may threaten to undermine perceptions of professional journalistic authority, so the likes of Wikipedia might call into question the public authority of the university academic or of the peer-reviewed publication.

Indeed it may further be noted that one man's freedom fighter is another man's rioter; and that therefore the perspectives of the media establishment upon civil unrest may in the end come down to how close (geographically and economically) that unrest comes to undermining their own established interests: that violence in Egypt, for example, represents to the UK media establishment (and, for that matter, to the UK academic establishment) legitimate political protest (and therefore the technologies which appeared to facilitate this violence are seen as liberating), while violence in London represents wanton criminality (and therefore those same technologies are in this case seen as morally corrupting and corrupt).

It might even be suggested that, in the case of the 2011 riots, what has really incensed certain elements of British journalism is the appropriation of certain new media technologies (once considered the province of a professional elite) not only by the general population but specifically by groups of British citizens perceived as representing an underclass (those so-called 'Chavs') – an underclass defined and denigrated by significant sections of the popular press in what might appear to be little more than a populist alternative to what has become a socially unacceptable level of ethnic prejudice. Just as the underclass have appropriated Burberry (once a symbol of moneyed power) they have now appropriated BlackBerry (a similar symbol): and, as the British press was wont to emphasize, they specifically *stole* these phones during their rioting sprees. (In a crimino-logical chicken-and-egg paradox, it remains unclear which came first: the looting of the phone or the use of the phone to organize the looting.)

This perceived appropriation of the technological tools of interactivity has further provoked the exclusion of this so-called underclass from the processes of societal interaction. It therefore perhaps seems worth reflecting for a moment upon the sociologically spurious nature of the underclass thus imagined and thereby engendered.

Blackberry is the New Burberry

The UK's 2011 riots kicked off on the weekend of 6–7 August. On 8 August 2011 the *Daily Mail* – the self-styled voice of Middle England – a ran a story headlined 'Tweeters fan the flames of hatred'. The following day the *Mail* announced that 'Young looters coordinate raids via Twitter and BlackBerry'. The next day another *Mail* headline denounced the 'BlackBerry ringleaders'. The day after that, a 1,400-word feature article by Tanith Carey asked 'Why do people become so vile online?' Carey conjectured in this article that 'the downside of the free speech offered by blogs, Twitter and social networking is that it has created a generation of narcissists obsessed with their own opinions.'

Could it then be that western societies are on the verge of defining socio-political class in terms of the perceived appropriateness of specific demographic and economic groupings' uses of new media technologies? Are we witnessing the evolution (or the invention) of a new mode of quasi-ethnicity determined upon the employment of mobile telephones and social networking sites? Would this situation indeed prove any more or less absurd than any other form of social discrimination?

On 12 August 2011 the historian David Starkey notoriously articulated what has been perceived by some commentators as a new brand of racism, a mode of discrimination based on a redefinition of race by cultural association. Speaking on BBC2's *Newsnight* programme in response to the series of riots which had engulfed British inner city areas earlier that month, Starkey argued that 'the whites have become black. A particular sort of violent, destructive, nihilistic, gangster culture has become the fashion. And black and white [...] operate in this language together.'

This quasi-racism, one without a rationality or fixedness of object, seems to expose precisely what racism is: a pre-judging of the other not because they are otherly but because they are designated as otherly. As Slavoj Žižek (1989: 99) has argued, a Jew, for example, is 'in the last resort one who is stigmatized with the signifier "Jew"' – and Starkey's perspective echoes this phenomenon. The assertion that white is the new black rests on the self-justifying presumption that black is black – indeed that black is not only *a priori* otherly but also meretricious and morally corrupt and corrupting – in Frantz Fanon's terms, 'not simply [...] lacking in values [but] the negation of values' (Fanon 1990: 32). It is not so much that white is the new black as it seems that prejudice against a dispossessed minority (only vaguely and spuriously defined by its ethnic profile; one just as easily defined by its new media profile) is the new (and apparently the acceptable face of, or the alternative to) racism – in that racism is as ill-defined as race: and the exposure of its own irrationality has led racism to modify (and to blur) the focus of its xenophobic prejudice.

And here, of course, in Britain in the summer of 2011, we have the perfectly ill-defined object of this perfectly ill-defined prejudice: an irrationality of reaction against an irrationality of otherliness. As one Labour MP pointed out, 'the polarisation is not between black and white. It is between those who have a stake in society and those who do not' (cited in Benyon 2012, 14). Indeed one might add that the polarization is specifically between those who are perceived by the political and media superstructure as having the right to a stake in society and to the technological capital necessary for such a stake, and those who are not. As we can no longer define the other on grounds of ethnicity (as it does not make sense to – and is not acceptable to), we have now returned to that perfectly meaningless notion of 'class' (perfect because meaningless, unassailable because beyond the rational) – in this particular historical instance, of what one tabloid newspaper journalist depicted as 'a feckless criminal underclass' (Wilson 2011), these 'feckless youths' (*Daily Mail*, 10 August 2011) – the spawn of 'feckless parents' (*The Sun*, 11 August 2011) – whatever the feck that might actually mean. (According to the 1972 *Supplement to the Oxford English Dictionary* to 'feck' is to steal. It seems that these people were hardly 'feckless' in this sense; but one suspects that morally outraged tabloid editors could not quite bring themselves to refer to these 'fecking' youths.)

The rapper-songwriter-actor known as Plan B (real name: Benjamin Paul Ballance-Drew) wrote of the UK riots, in an opinion piece for *The Sun* newspaper on 10 August 2011, that 'they're not attacking the Government. If this was about high taxes, things costing too much money in this country, why are you attacking the working class, the retail shops?' The Slovenian cultural philosopher Slavoj Žižek (2008: 64–65) has similarly written of the irrationality of the riots which overtook the suburbs of Paris in the autumn of 2005:

> there was *no* programme behind the burning of the Paris suburbs [...] opposition to the system cannot articulate itself in the guise of a realistic alternative [...] but only take the shape of a meaningless outburst [...] The cars burned and the schools torched were not those of richer neighbourhoods. They were part of the hard-won acquisitions of the very strata from which the protesters originated.

Žižek (2008: 70) also sees this anti-rationalistic outpouring of violence in the 'psychotic-delirious-incestuous' phenomenon of terroristic fundamentalism – which is again a phenomenon which has appropriated the new media technologies of western hegemony – much to the chagrin of that hegemony: as, for example, the *Daily Mail* (drawing a direct comparison between UK rioters and Islamic terrorists) pointed out on 9 August 2011, 'several countries [...] have complained that BlackBerry messages [...] may be used by terrorists.'

The division of the self and other – or of superstructure and underclass – has never been anything more than an ideological alibi, a myth of power (and in this case a myth rooted in the entrenchment of the institutions and technologies of political and media power). In the end, despite their furious denials, these illusions of moral, social or 'ethnic' distinction fail to make any plausible kind of difference. In these terms, xenophobic aggression can be seen as the result not of discrimination but of the failure of discrimination, insofar as xenophobia does not in fact fear difference: xenophobia fantasizes about difference and fears (and therefore denies the fact of) similarity. It is the underclass's appropriation of our symbols and tools of wealth and power which provokes this aggression: when, for example, those perceived as culturally illiterate join the ranks of the Twitterati. It is because *they* are threatening to become too much like *us*.

There is a clear paradox in all of this. The historical persistence of anti-Semitic aggression, for example, attempts to conceal its second greatest secret: the secret that the secret of the secret of the Jews is that the Jews have no secret. Its greatest secret is that it of course already knows this. A similar condition applies both to contemporary Islamic fundamentalists and to contemporary Islamophobes. One might be forgiven for imagining a massive and unspoken conspiracy between those two polar extremes (or at least a mutual and self-sustaining ideological delusion), a repressed complicity in the denial of the fact that those two extremes do not in fact represent polar opposites at all. Andrew Osbourne (2011) neatly summed up the fear engendered by this uncanny similarity of otherness when he wrote in *The Daily Telegraph* of the Islamic fundamentalist Vitaly Razdobudko's suicide bombing of Russia's busiest airport in January 2011: 'Staring out from the front pages of their newspapers this weekend is not the usual dark-skinned, heavily-bearded Islamist terrorist they have come to expect and fear but an ethnic Russian who looks like millions of Russians' brothers, sons or husbands.' Long gone, it seems, are the reassuring times in which the villains twirled their absurd moustaches and dressed in conveniently sinister hats and capes.

The enemy within is one which is, in Umberto Eco's words, 'at the same time inside and outside' (Eco 1995). That is what really scares us. That is why so much anti-Semitism and homophobia just won't go away – because Jews and gays aren't always obvious (and that is why, for example, male homosexuality seems more acceptable to a homophobic society if it is overtly camp or effeminate); and that is why Islamophobia all too often focuses on the veils (we cannot see who they are; they could be anyone – any one of us) and on the converts (they don't look like Muslims; we can't tell if they are; we might become like that too). And that is why the right-wing press have found some reassurance in the potential at once for new media technologies to conceal and reveal the identities of these subversive elements. As the *Daily Mail* announced on 9 August 2011, 'intelligence experts are scouring public messages for evidence of those behind the mayhem' – and on the following day: 'government spies have been drafted in to track riot ringleaders who have been using encrypted instant messages on their BlackBerry smartphones to avoid detection.' In other words, says the voice of Middle England: we know what you did last summer.

So if the appropriation of these communications technologies really so threatens the institutions of established power that this appropriation has become a determinant in the outraged denigration of a disempowered section of society, then might these technologies in fact eventually offer to empower the underclass and overthrow the entrenched structures of power? Are the technologies empowering in themselves: or is it the case that, when content (or information) is king, these technologies (which are in themselves no more than highly efficient but otherwise empty vessels) can only offer us a reiteration of that old computing cliché: *garbage in, garbage out*?

The Knowledge Economy

Might it be argued that Web 2.0's emphasis on user-generated content which underpins what Andrew Keen has described as a 'cult of the amateur' represents a radical (and eventually narcissistic and solipsistic) individualism which undermines the stability of hierarchical structures *per se* (shifting order towards chaos: dictatorship towards democracy, democracy towards anarchy)? What the Internet promised us, in our cyberutopian dreams, was the possibility of a new and liberating consensus based upon dialogue and mutual understanding; but what it has delivered – from 9/11 to the

Arab Spring, from anti-capitalist protests to cyberbullying, from Greek
riots to London looting – is (as Plan B and Slavoj Žižek incongruously
agree) a contentless violence born out of the frustration of anger without
an agenda (without a realisable and workable agenda, however utopian;
that is, without a shared vision). No new global enlightenment has as yet
been bred from a digital public sphere; rather, the diverse and directionless
voices of these narrowcast private spheres have been unable or unwilling
to listen to each other in the chaos of these irresponsive monologues. The
promise that new media technologies might sponsor a liberation from
totalitarian or authoritarian regimes is true only insofar as liberalization
is the first stage of anarchization; and if this clamour of unlistening voices
might deconstruct modes of oppressive authority and move oppressed
societies towards forms of liberal democracy, then it also moves liberal
democracy towards the anarchy of unstructured and anonymous (and
therefore unaccountable) power. Or rather it only does so until structures
of power (either traditional or new) reassert themselves – in omniscient
hierarchies which are more anonymous, less visible and therefore even
less accountable (although possibly more demagogic) than ever before,
and which by claiming to offer empowerment in fact diminish popular
resistance to their power.

The deployment of digital media technologies towards a project which
calls itself populist or even democratizing is not, then, necessarily condu-
cive to the progress of democracy as some might understand it. One might
suppose that democracy is a term which is often invoked in defence of its
opposite: after all, how many totalitarian dictatorships have over the last half
century or so described themselves as democratic republics? Co-founder of
Wikipedia Larry Sanger has commented that there is clearly a problematic
ambiguity in the ways in which we speak of the processes of democratization: 'it depends on what you mean by "democratization". Does the question concern progress toward progressive political ideals, for example, or
toward basic and literal democracy (which isn't much of a concern in our
countries), or toward a greater degree of "empowerment" of the formerly
silent individual in the marketplace of ideas – or what?' This nebulous
term is repeatedly used to legitimize a variety of very different positions.
Is it, for example, democratic that new media give their users a voice, even

though it may be that nobody listens? Is it democratic to promote self-expression without any attempt to evaluate (and therefore to value) what is expressed? One might suppose that what one would like to mean by democratization represents the general empowerment of the individual – and one might therein argue that we are empowered not only by access to the tools of self-expression, but also by access to sufficient knowledge and information to allow us to form and express our judgments through critical and contextualized reflection, which would then promote the recognition of those positions through dialogue and informed debate. In other words, what we may mean by 'democratization' may bring us back towards the promotion of what Jürgen Habermas would call a public sphere – an open, dynamic and transparent forum for public dialogue.

The British-born Silicon Valley entrepreneur turned cybersceptical Internet apostate Andrew Keen has, however, supposed that in the world of the Web 'the more people use the word "transparency" the more secret they tend to be.' Keen has also argued the economic case that many of the platforms of Web 2.0 – and here he has specifically cited Wikipedia – may not only be seen as textually undemocratic (despite all their claims to the contrary) but may also be viewed as exploiting the labour of their users – who receive neither recompense nor recognition: 'I'm deeply hostile to the idea of giving one's labour for free without getting any benefit. And when you give away your labour anonymously, that's really problematic.' From this perspective we might see Wikipedia as the bourgeois workhouse of the twenty-first century.

Keen has suggested that Web 2.0 has little to do with democratization. He has argued that, when cyberenthusiasts speak of the Internet's powers of democratization, they appear to be suggesting that 'it flattens the playing field – everyone has the same voice – it breaks down the hierarchies, the supposed oligarchies of the old media world.' Keen has supposed that 'words like democracy get thrown around all the time, but not everyone agrees about the ideal of democracy.' He has added that 'the people who once had a monopoly on the information economy no longer have that – now anyone can be an information mogul.' Yet Keen has suggested that these new information moguls have themselves formed 'the Internet's digital elite' and has criticized this elite's hypocrisy: 'one of the annoying things is that they continue to speak in the voice of the people, but they are the elite. The counter-culture has become the establishment. The new elite has seized power in the name of the people.' Keen has stressed that this situation represents 'a new challenge to representative democracy.'

Keen has proposed that 'this idea of democratization' is 'a very American idea.' He has argued that

> most of it stems from America – a western ideal of regeneration, of starting again, of reinvention, of expansion. It reveals the very ill-informed idealism of many American progressives. In many ways what they've done is to replace the myth of America with the myth of the Internet. It's also no coincidence that there's a connection between born-again religiosity and the Internet. That utopianism extends between religion, the West and the Internet. It shows a fundamental ignorance of history – that these things have been done before. They think that you can begin again, that history has nothing to offer us – but it's always ended in bloodshed. These people – these technologists – have no education in history. They know nothing about the past. I think a lot of them mean well, but I don't think they know what they're doing.

At the end of George Orwell's novel *Animal Farm* the leaders of the revolution – the pigs – join forces with the human elite. Orwell (1951: 120) famously wrote that the other animals 'looked from pig to man, and from man to pig, and from pig to man again; but already it was impossible to say which was which.' The conclusion to Orwell's fable offers two messages which may still seem pertinent to the Internet generation. The first is obvious enough: that, as power corrupts, the revolutionary counter-culture itself develops an elitist hierarchy.

In 1967 the American poet Richard Brautigan imagined a counter-cultural cybertopia in his poem 'All Watched Over by Machines of Loving Grace'. In 2011 BBC television screened filmmaker Adam Curtis's documentary series of the same title. Curtis suggests the failure of the counter-cultural project in a way which reflects Orwell's metaphor: 'we live in a dream world created by the machines where we are all connected and we are all free [...] we dreamed the systems could [...] create a perfect balance [...] without [...] the old hierarchies of power. But power hasn't gone away. It never does.'

But we can perhaps see a second level of meaning in Orwell's image: the tragedy is not only that pig and man are objectively indistinguishable; the tragedy is that their audience can no longer distinguish between them. And so we view the new elite as our liberators and see liberation in the new hierarchies; we mistake our unheard vocality for power; we accept the rumours, Chinese whispers and more malicious fabrications which arrive through our machines in lieu of the tested, grounded, validated knowledge which would offer the only actual possibility of our empowerment.

CHAPTER 7

Waiting for the Great Leap Forwards

That truly remarkable book, that imaginary prototype of new media inter-activity, that fantastical forerunner of the populist electronic encyclopae-dia, Douglas Adams's *Hitchhiker's Guide to the Galaxy* famously noted in the late 1970s that the planet Earth was 'mostly harmless'. While this volume has generally characterized the more ambitious claims of liberat-ing interactivity advanced by the advocates of new media technologies as possibly somewhat worse than harmless, it should again be emphasized that the central role which these technologies play in our lives might – if the scope of their impact can be properly understood – continue for the most part to facilitate our social and political interactions in ways aligned to a consensual notion of progress towards the enhancement of the public good, and might propel media and policy agendas beyond the reactionary constraints and hierarchies of entrenched institutions steeped in elitist and exclusionary structures of power.

Yet, insofar as it has promised swift, easy and ultimately unsustainable solutions to our social, political and existential woes, the new media revo-lution has thus far provided a comforting but fundamentally counterpro-ductive distraction from these issues rather than the practical facilitation of the development of progressive agendas designed to address them – a function for which it clearly retains some significant potential. This chap-ter will therefore briefly explore, through the fields of media, politics and philosophy, whether and how these interactive technologies and formats might still have a role to play in the discovery of viable solutions to these ongoing problems.

Stop Press: Press to Continue

On 10 July 2011 Britain's biggest selling newspaper *The News of the World* published its final edition – after 168 years of publication. The newspaper closed as a result of a scandal which had involved the illegal hacking of the mobile telephones of a large number of public figures and private citizens, and which called into question the legitimacy of the newspaper's (and its multinational corporation's) relationship with government and the police. In effect, the newspaper was brought down by fears that in an age of global multimedia communications its megalithic powerbase had undermined democratic and transparent relations with the traditional institutions of politics and the law. This has proven to be perhaps the most significant example to date of an ongoing collapse of trust in the relationship between media and democracy: a crisis which led to the 2011 launch of Lord Justice Leveson's public inquiry into the conduct of the British press and which prompted the UK's parliamentary Culture Committee to state in May 2012 that Rupert Murdoch was not a fit person to run News Corporation. And yet, despite Mr Murdoch's dramatically public transformation from a Hyperion into a rather sorry-looking satyr, what alternatives do we have?

In his exceptional study of *Online Journalism* Jim Hall (2001) proposed – citing cases in the former Yugoslavia, and anticipating those since seen in Iraq – that the weblog might offer a significant alternative source of information and reportage to traditional news outlets. However, Rupert Murdoch's somewhat ill-judged purchase of the Myspace blogging platform in 2006 appeared to demonstrate that older media empires were willing to strike back to assimilate these forms of unauthorized expression; and the escalating levels of corporate power and control displayed by new media empires, and the questionable quality and relevance of the vast majority of independent user-generated content, have somewhat diluted hopes for the impact of such de-institutionalized media sources.

The underground blogger, of course, like the software entrepreneur starting up in his parents' garage, is swiftly professionalized, institutionalized and corporatized by any hint of popular success. When, for example, one sees Salam Pax, the Baghdad Blogger, filing a series of reports from Iraq for BBC Television's *Newsnight*, one is forced to wonder how unmediated his reportage now is. Doesn't such success defeat the weblog's end, by thrusting it into the mainstream? And when *The Huffington Post* – launched in 2005 as an independent news aggregator and outlet for progressive opinion – launches international editions in Canada and Western Europe, is bought by AOL and wins a Pulitzer Prize, it is difficult to see how alternative it has managed to remain. Indeed, as Leaning (2011: 98) suggests, blogs may more usefully be seen as part of an interacting mesh of media forms rather than as separate or alternative forms in themselves. They do not, as such, represent an alternative to entrenched institutional structures of information and communication but an extension of those structures.

The well-known British political blogger Iain Dale has suggested that the political weblog has promoted levels of political engagement among younger voters in the UK: 'I think blogs have engaged people who might otherwise not have become involved in the political process, especially among the 18–25 age group. I know this from dozens of emails I receive from young people, who may be turned off engagement with political parties, but want to involve themselves in the democratic process. They see blogs as a medium for doing this.' Although Dale does not see the blog replacing or even seriously rivalling traditional news services, he acknowledges the weblog's increasing role in setting the news agenda: 'I don't believe that blogs will replace old media, but there's no doubt that new media has had a major impact on the old media. They have provided the old media with new voices to report and engage with and use as independent pundits and commentators.' Dale also recognizes that the weblog phenomenon does not necessarily represent an alternative media form which promises the democratic redistribution of political influence away from established voices: 'In some ways mainstream journalists have already colonized the blogosphere so blogs have, to an extent, already become institutionalized. This is inevitable and is neither a good nor a bad thing. It's just a fact.' Modes of resistance to institutionalized power themselves become institutionalized and entrenched: indeed, Dale himself is now something of a recognized figure of the media establishment.

The writing has for some time been on the wall for the truly independent popular news-related blog. For the most part, we have not performed a massive shift in our choice of sources of news. The Internet has provided a different medium through which to access these sources, yet we continue to return to the usual authorities with their same-old stories: *The Guardian*, *The Times*, *The New York Times*, *The Sun*, Reuters, the BBC, CNN. Our recourse to these sources may not merely result from an innate conservatism or inertia: it may be because we understand that 'news' is not merely what has happened in the world today, but what has been deemed to be significant or newsworthy by forms of recognized authority, and that which thereby becomes part of a shared, global, gate-kept experience. It may be also that – despite allegations of telephone hacking and other inappropriate practices conducted by certain elements within the mainstream media – there remains a lesser public distrust of these organization than of the Internet's less authoritative portals. In February 2008 the *Drudge Report*'s decision to give wide publicity to the then embargoed news of Harry Wales's sojourn in Afghanistan prompted some moral discomfort; and this is perhaps why we prefer to read the latest from Wikileaks through the ethical and informational filter of *The Guardian* newspaper.

The websites of major news organizations once commonly offered their readers links to alternative news sites and blogs; this practice is now somewhat rarer. Instead (and in line with these processes or re-institutionalization) they have begun to offer their own informal news(b)logs – such as BBC Political Editor Nick Robinson's blog on the *BBC News* website. Robinson's blog has, at times, displayed a telling antagonism towards news blogs produced by those outside the institutions of professional journalism. In an entry on political bloggers' vilification of the UK's then Deputy Prime Minister John Prescott, Robinson (2006) argued that 'this is another example of some blogs trying to make the political weather.' Robinson also cited the uses of weblogs by the Conservative Iain Dale, and by *Swift Boat Veterans for Truth*, to tarnish the reputations of Cherie Blair and John Kerry respectively. Robinson distinguished between the professional conduct of journalists and the unregulated activities of bloggers, and proposed that the decision to report unsubstantiated allegations is not a mark of good journalism but of political bias.

In August 2006 Iain Dale publicly defended the rights to 'freedom of speech' of his fellow right-wing blogger Inigo Wilson (Roberts 2006) when the latter was suspended from his job as a result of a poorly written, ill-informed and racially offensive diatribe he posted on the right-wing website ConservativeHome.com. In such a context one might appreciate Nick Robinson's attempt to champion a more formal mode of political writing which eschews blatant prejudice and polemic: yet underlying his argument is the assumption that professional journalism can somehow attain the impossible position of ideological neutrality – and that it thus alone has the right to conjure political storms.

The rise of the independent blogosphere has corresponded with a certain amount of media hype surrounding a phenomenon which has become known as citizen journalism. The main problem with the idea of citizen journalism is, however, to put it bluntly, this: citizen journalism does not really exist. The successful citizen journalist tends to become a professional journalist; and, for the most part, what has been hailed as citizen journalism constitutes either poorly informed and generally unread opinion pieces which languish obscurely on individuals' personal blogs, or the capturing of photographic images or video footage on mobile phones (and sometimes the posting of this raw material onto public websites) which is then used (as witness testimony has always been used) by professional journalists. While this latter development has, as Jon Silverman (2011) observes, proved a useful resource in the news-gathering process, it hardly represents a culture shift in the way that the majority of people receive their daily news. Few news-consumers directly trawl Blogger, Flickr or YouTube for their news: this material is usually accessed from the services of traditional journalistic outlets, after it has undergone the standard processes of professional mediation (selection, editing and interpretation) – and this may then be accessed by news-consumers from these institutional sites, albeit increasingly via recommendations on such sites as Twitter and Facebook. Nearly 20 per cent of online news consumers arrive at stories via social media, and news institutions are increasingly collaborating with such portals (Ashton 2012). Thus the public may be at the gates, but the gates are still being kept by the traditional authorities.

Things Can Only Get Better

Despite Iain Dale's optimism as to the potential of new media platforms to encourage political participation, increasing numbers of commentators have appeared concerned by an apparent public apathy in relation to participatory democracy. As this volume has suggested, the rise of reality television, video games and online social networking may divorce citizens from the desire for social and political agency which is traditionally viewed as essential to democratic citizenship; and, just as citizens appear increasingly distanced from the political arena, so that arena itself seems to have diminished, or to have become detached from what western civilization has for centuries defined as politics, and hence from the history of politics itself.

The parliamentary election which took place in the United Kingdom in May 2010 had been heralded by many as offering a return of politics to the people. It is certainly the case that the election witnessed a modest reversal in the general decline in voter turnout – up to 65.1 per cent from 61.4 per cent at the previous general election of 2005. This may have been the consequence not so much of the uses of new media technologies and formats during the campaign as of the fact that the result was far from being a foregone conclusion. (Indeed the results of Britain's 1992 election, which had witnessed a 77.7 per cent turnout, had been similarly uncertain.) By the local elections of May 2012 turnout had fallen to only 32 per cent, the lowest figure reported in more than a decade. Support for the government that had been brought to power by the 2010 election (a government with a questionable mandate secured out of an election in which no party had secured a majority of parliamentary seats – let alone, of course, an outright majority of the popular vote) had also considerably waned.

Nevertheless, on 6 May, the day of the 2010 general election, the front page headline of *The Independent* newspaper had hailed that event as 'the people's election'. It argued that through the country's first ever series of televised debates between the leaders of the main political parties, and as a result of online activities by individual bloggers and posters which challenged and subverted the propaganda of the established parties, influence over the political agenda had shifted into the hands of the British people. The results of that election, however, did not demonstrate any significant shift in public opinion away from the two largest political parties; the smaller parties and independent candidates did not make the expected electoral breakthroughs which might have diminished the Labour and Conservative stranglehold over the British political system that had endured for the best part of a century. Independent bloggers remain, of course, for the most part unread (how could they all be read?), while the World Wide Web continues to spew forth such democratic delights as the Labour Party YouTube channel (launched by Tony Blair in April 2007), a YouTube version of Prime Minister's Questions (launched by Gordon Brown in May 2008), and, perhaps most disturbing of all, the video weblog launched by Conservative Party Leader (and future Prime Minister) David Cameron in September 2006, Webcameron, on which one could view the great man pontificating on the issues of the day from the comfort of his luxury kitchen. In short, new media forms – from Estonian e-voting through to British blogs – continue, as ever, to be appropriated as tools for the reactionary reinforcement of entrenched power.

In May 2010 the polling organization YouGov reported that 29 per cent of UK viewers believed that the televised debates between the leaders of Britain's three main political parties (a series of three debates run for the first time during this election campaign) had been the 'most interesting programmes on TV' over that period. The science fiction series *Doctor Who* scored only 17 per cent – while only ten per cent of those questioned considered the popular reality television series *Britain's Got Talent* the most interesting show then on. Although it is perhaps clear that the heightened interest in the political process may have been more closely related to the closeness of the parties' standing in the polls (which eventually resulted in a hung parliament), it has been suggested by some commentators that the popularity of these televised debates was responsible for the increase in voter turnout at the subsequent election.

Figures, however, released by the Broadcasters' Audience Research Board (BARB) do not necessarily demonstrate such a healthy public interest in democracy. While the YouGov statistics show that those who watched the leaders' debates found them interesting, overall viewing figures suggest that these debates generally remained less popular than reality and fantasy TV. The first week of the debates saw *Britain's Got Talent* achieve 11.87 million viewers; *Doctor Who* 7.82 million and the election debate (screened by terrestrial broadcaster ITV) 9.68 million. Although lagging behind Simon Cowell's reality television show, the debate did gain a respectable audience share: this was, after all, the first time such a debate had ever been screened in the UK. The following week, the debate moved to the satellite broadcaster Sky News: while *Britain's Got Talent* took 11.45 million viewers and *Doctor Who* managed 8.13 million, the party leaders' debate had plummeted to 2.21 million. Back on terrestrial television, the third and final debate in the series – broadcast on BBC One – managed to increase its ratings to 7.43 million viewers. That week, however, *Doctor Who* attracted 8.02 million viewers and *Britain's Got Talent* took 11.72 million. Although the first debate had been the second most popular programme of the week (across all British television channels), the third debate was only the sixth most watched programme of the week on BBC One and was also beaten in the BARB ratings by eight programmes on ITV. (The second debate, on Sky News, did not feature in the cross-channel top 50 programmes of that week – indeed BBC One and ITV each managed more than 30 programmes with higher ratings that week.) The cumulative viewing figures for the three debates totalled 19.32 million – giving a weekly average of 6.44 million viewers. Over that three-week period *Doctor Who*'s ratings averaged 7.99 million, while *Britain's Got Talent* averaged 11.68 million. In other words, the average viewing figures for the leaders' debates were just slightly over half (55 per cent) of the ratings for *Britain's Got Talent* over the same period.

Popular entertainment once again trounced national politics. This situation continued in the aftermath of the election itself as the political parties scrabbled to forge a deal for coalition power. On 14 May 2010 the *Daily Mail* newspaper reported that more than a thousand people had complained when (three days earlier) the BBC had chosen to postpone episodes of the soap operas *EastEnders* and *Holby City* in order to broadcast live coverage of the departing Prime Minister Gordon Brown's resignation speech from Number 10 Downing Street. As the *Mail* pointed out, 'despite historic changes in the real world, it seems many would have preferred to have watched the fictional goings on around Walford's market stalls and in Holby's hospital.'

On 12 May 2010 the *BBC News* website pointed out that an average of 8.9 million people had watched the BBC's coverage of Gordon Brown's resignation the previous evening, although BARB figures (published two weeks later) put this at 8.73 million. According to statistics published by BARB, however, the previous Tuesday's episode of the soap opera *EastEnders* had attracted a slightly higher audience of 9.58 million viewers. Beaten in popularity by mass media entertainment, politics comes to take on the characteristics of that entertainment: indeed the media focus on the subsequent campaign for the election of Gordon Brown's replacement as leader of the Labour Party became what candidate Ed Balls described in August 2010 as a 'daily soap opera.'

Even if the interest in the UK's televised debates had proven a sustainable strategy for a restoration of public involvement in political processes, some have questioned whether the telegenic status of the candidates is necessarily the best way to judge political integrity and ability. On 2 May 2010 *The Observer* newspaper quoted a last-ditch attempt by the then Prime Minister Gordon Brown to counter the prevailing shift of political culture into the formats of popular television entertainment: 'We're talking about the future of our country. We're not talking about who's going to be the next presenter of a TV game show. We're talking about the future of our economy.' Four days later, Mr Brown lost his parliamentary majority and five days after that his government was swept from power. Ironically, Brown had repeatedly pledged his support for reality television over the previous two years – something which the *BBC News* website emphasized just two days before the election (and just two days after Brown's comments in *The Observer*): 'Gordon Brown is a huge fan of the Simon Cowell brand of reality TV. He has spoken several times about his love of ITV's *X Factor* and *Britain's Got Talent*.' It appears the British public were unwilling to believe his belated warning as to the game-show dumbing-down of UK politics, or perhaps remained distrustful as to the sincerity of that warning.

One might recall in this context Marshall McLuhan's comments on the U.S. presidential election debates between Richard Nixon and John F. Kennedy televised in 1960: 'TV would inevitably be a disaster for a sharp intense image like Nixon's, and a boon for the blurry, shaggy texture of Kennedy' (McLuhan 2001: 359–360). McLuhan (2001: 361) went on to detail Kennedy's appeal: 'Kennedy did not look like a rich man or a politician. He could have been anything from a grocer or a professor to a football coach. He was not too precise or too ready of speech in such a way as to spoil his pleasantly tweedy blur of countenance and outline.' Kennedy was a mediated everyman, for McLuhan an image perfect for, and created by, TV. He was, in other words, the archetypal game show host. Gordon Brown's failure to woo the voters may be attributed in part at least to his lack of such telegenic charisma – as the former Labour Party leader (and ostensible ally) Neil Kinnock had told the BBC in April 2010, 'Gordon has got a radio face.'

It is, of course, a cliché of media history to suggest that while Kennedy won the television debate, radio listeners tended to feel that Richard Nixon had won the substantive argument. Kennedy's election victory thus demonstrated the supremacy of television over the older medium, and thereby of style over substance. Yet the British electorate's response to Brown's appeal against the celebritization of politics might be read in another way. Although the public did not come out in droves in support of Brown (the then Labour leader and self-confessed *X Factor* fan), nor did they afford a ringing endorsement either to David Cameron (Conservative Party leader and former Director of Corporate Affairs at the UK media company Carlton Communications) or to Nick Clegg (the Liberal Democrats' leader, a former journalist and political lobbyist at whom Brown's 'game show host' jibe had primarily been aimed). Indeed, despite his apparent success in the televised debate and the outbreak of what the media had dubbed *Cleggmania*, Clegg's party lost seats at the subsequent election.

It might therefore be hoped by some that the tensions between traditional and contemporary models for the mediation of the political process may be leading to a situation which neither reinforces nor outrightly rejects those tensions but reconciles them within a new model. It was suggested, for example, by *The Independent* newspaper the day before 2010s election that the expected hung parliament resulting from voter dissatisfaction with the nation's traditional political structures might lead to the 'historic opportunity' of significant parliamentary and political reform: the development of a new mode of democracy, one more appropriate to the new media age. However, it remains to be seen whether such dreams can be realized. It was certainly the case that the electoral reform envisaged by *The Independent* failed to engage the imagination of the British public when it was put to a referendum a year after the election, on 5 May 2011.

This referendum on electoral reform had been agreed a year in advance as part of the Conservatives and Liberal Democrats' negotiations on forming a coalition government. On Wednesday 12 May 2010 those two parties had formally launched the first coalition government in the UK since the Second World War. This response to the nation's half-hearted election of a hung parliament was described by the incoming Prime Minister David Cameron as the beginning of a 'new politics' – 'a historic and seismic shift in our political leadership.' Indeed, the coalition was heralded by a front page headline in *The Sun* newspaper that morning as the beginning of a 'Dave new world'.

There was clearly a desire among political leaders to signal to a dissatisfied electorate the development of a new governmental paradigm, but it seems in retrospect that this declared position did not so much demonstrate a will to implement a radically reformed political covenant as it represented a public relations stunt designed to capture the mood of the nation and to sustain an obsolescent structure of governance. On 14 May 2010, just two days into this era of new politics, it emerged that the first major political reform proposed by this coalition would be to change the proportion of MPs required to dissolve a parliament in a vote of no confidence against the government from a simple majority to 55 per cent, thus ensuring that even if the coalition were to collapse the Conservative administration could remain in power. For many Members of Parliament, both on the left and the right, this plan immediately called into question the democratic and progressive credentials of David Cameron's new model. Cameron's Deputy Prime Minister, Liberal Democrat leader Nick Clegg, may have spoken of 'a wholesale, big bang approach to political reform' during his first major speech in that office on 19 May 2010, but the radical reform of Britain's political system has not been apparent to all commentators. As Ann Treneman wrote in *The Times* on 19 May 2010, as the new parliament sat for the first time:

> The New Politics began not with a bang but a seating plan. Presentation, presentation, presentation! That is the mantra of our new rulers. The huge question of who would sit next to whom on the front bench could not be left to chance [...] Other than the seating plan, the New Politics appears to carry on exactly like the old.

On 1 July 2010 the UK's Liberal-Conservative coalition government announced the launch of the 'Your Freedom' website, a platform from which – in the words of a *BBC News* report – the public would be able to 'nominate laws and regulations they would like to see abolished [and] also be able to propose ways to reduce bureaucracy.' Although this was reported by the BBC as being advanced as the biggest online policy-sourcing initiative by 'any government ever', one might note its similarities to the similar sites launched by the Estonian government over the previous decade, and the failure of those sites to sponsor meaningful public debate – and their propensity to engage and empower a socio-economic and educational elite and to invoke and promote policies designed to reduce the influence and cost of the public sector.

One recalls Marina Hyde's suggestion in *The Guardian* newspaper (in January 2010) that online exercises in public consultation suggest a lack of actual coherent policies, Bynum and Rogerson's argument (2004a: 6) that such consultations may result in policy development coming to reflect knee-jerk public responses, and Needham's concern (2004: 65) that such populist approaches might allow governments a mandate to sideline parliamentary processes. The coalition's Deputy Prime Minister Nick Clegg appeared on the BBC's *Breakfast* news programme on 1 July 2010 to propose, rather surprisingly, that this was, in some way, a comic exercise in the recognition of legislative absurdity: 'This could release a sense of fun as people think of silly rules that need to be scrapped.' Clegg added: 'I've just discovered for instance, would you believe it, that there's still an old law in the statute book that says it's an offence if you don't report a grey squirrel in your own back garden.' Clegg's example recalls the absurdity of those similar exercises in online consultation performed in north-eastern Europe. One might therefore imagine that this initiative represents nothing more than a populist veneer of democratic consultation which offers the electorate an illusion of agency rather than a significant process of empowerment.

Indeed, after a month of this crowdsourcing policy initiative *The Daily Telegraph* reported, on 3 August 2010, that 'despite tens of thousands of public responses sent to various government departments, not one has shown a willingness to amend policy.' The *Telegraph* noted that, although the government had received responses in areas as diverse as health provision, pensions and taxation, transport and wildlife policy, each Whitehall department's official response appeared to interpret the public comments as 'an endorsement of existing policy.' Like the New Labour project which preceded it, the coalition's New Politics has lacked depth and substance: there was, as Jean Baudrillard would have said, only the image.

In August 2011 the same British government established a website to accept e-petitions. An attempt under the previous Labour administration had – according to Scott Wright (2011) – been hijacked by a relatively small number of serial petitioners. The petitions on this site had, for example, included one signed by 50,000 people in 2008 suggesting that motoring television present Jeremy Clarkson should be appointed Prime Minister. But in the second decade of the twenty-first century the new coalition government was of course promising something new.

On the day of its 2011 launch, the coalition's latest website, which promised that any petition receiving more than 100,000 signatures would be debated in Parliament, had already fallen foul of eccentric minority interests. As *The Daily Telegraph* reported on 4 August 2011, 'right wing Internet bloggers have been collecting signatures for several days calling for the reintroduction of the death penalty.' Later that day, the BBC noted that the return of the death penalty headed what it described as 'the list of demands.' The list also included the nation's departure from the European Union and limiting prison food to bread and water. The following week, however, the BBC reported that, in response to the series of riots which had swept the nation over the previous few days, the most signed e-petition on the website was one which called for convicted rioters to lose benefit payments. The petition, which had swiftly gained the necessary 100,000 signatories, specifically proposed that 'no taxpayer should have to contribute to those who have destroyed property, stolen from their community and shown a disregard for the country that provides for them.' One would assume that this rather blanket statement might suggest, somewhat beyond its original context, that not only social security benefits but also health-care, education and other public services (including prisons, courts and the police) should be denied to anyone convicted of committing crimes against property. As such, it might appear as something of a manifesto for the creation and perpetuation of a permanent underclass. This mode of poorly considered, ill-informed and therefore fundamentally undemo-cratic demagoguery seems to represent the politics of the mob and of knee-jerk hysteria rather than the reasoned debate of a public sphere. As a *Guardian* newspaper editorial of November 2011 cautioned, 'this risks becoming mob rule.'

In fact the British government went on to do its best to ignore this e-petition and its demands were sidelined in the parliamentary response to the riots and were never put forward to be enacted into legislation. This seems to demonstrate, yet again, that while new media may be useful in riling demagogic fervour to make the masses think their voices count, power elites continue to ignore those voices. Indeed, even such well-meaning online movements as the Invisible Children campaign for the arrest of the Ugandan guerrilla leader Joseph Kony have, as Harding (2012) points out, been criticized for their western-centred naïvety and their lack of useful impact – insofar as, as one online critic suggested, the humanitarian awareness of American college students does not represent a prerequisite for conflict resolution in Africa.

In August 2011, shortly after the success of the petition for rioters to lose their benefits, another e-petition appeared on the British government website – although it was removed the following month when someone in authority eventually got round to noticing it:

> All rioters and looters from the recent troubles in English cities should be banished to the Outer Hebrides for five years. This would be much, much cheaper than keeping them in expensive prisons, saving the taxpayer money. Five years of being forced to live in the Outer Hebrides with none of the comforts of English city living e.g. running water, electricity, decent food, culture and shopping, will put them on the straight and narrow, and frighten them not to riot or loot again. Many local people there look after sheep part-time, so they can earn a small amount of extra money looking after rioters and looters as well.

This satire cleverly subverted the demagoguery fostered by this illusion of democratic participation by exposing its innate absurdity. It is not the only voice of protest we have heard.

The Pursuit of Happiness

Even in today's society of the simulacrum – this Andy Warhol screen print of a celebritized civilization – there remain calls for a return to a politics of substantive issues and democratic debate. In June 2007, to cite a very public and appropriately absurd example, it was reported that MSNBC newsreader Mika Brzezinski had attempted to rip, shred and burn her script live on air in protest at being made to lead the bulletin on hotel heiress and reality television star Paris Hilton's release from jail on driving charges ahead of news items on the ongoing war in Iraq.

The nineteenth-century Danish philosopher Søren Kierkegaard argued that we can only counter despair through a recognition of absurdity. In his great work *Fear and Trembling* Kierkegaard famously recalled the story of Abraham and Isaac, the Biblical account of the father instructed by God to sacrifice his son, the man who meets that divine paradox with the paradoxical absurdity of his own faith. Kierkegaard (2006: 49) observes that the triumphant absurdity of Abraham's faith lies in the fact that he as a single individual becomes 'higher than the universal' – an absurdity mirrored for Kierkegaard in history's greatest absurdity, the incarnation of divinity as an ordinary individual human being (Kierkegaard 2009: 117).

Roland Barthes (1977: 134) witnesses something of similarly transcendental significance when he observes how weakness defeats strength in the Biblical story of Abraham's grandson Jacob's struggle with the angel. The utopian myth of new media interactivity has offered us something similarly – almost impossibly – absurd: the ordinary individual made equivalent in power and significance to the totality of society; the mobilization of the individual within effective processes of democratic participation;

the dialogical development of a socio-political agenda by consensus; the apotheosis of all ordinary individual human beings as avatars of social, cultural and political agency. This myth might be said to parallel another famous Biblical account of weakness beating strength, the story of David and Goliath – if, that is, instead of killing Goliath, David had invited him to sit down and develop a dynamic, interactive, pluralist and egalitarian process for the establishment of consensual decision-making, one in which all voices, big and small, carried equal and essential weight. This, in effect, is what evangelical cyberutopianism has promised us; and yet what has been delivered by so many of the political and commercial interests which have come to dominate new media has been the precise opposite of this, a mere sop to any popular desire for empowerment.

Theology aside, the myths of Abraham, Jacob and David represent the individual's arduous, risk-laden and almost impossible struggle against massive power as being both absurd and, at the same time, definitively human. To acknowledge our disempowerment (indeed, to acknowledge the absurdity of the possibility of our empowerment) may eventually afford our only possibility of empowerment. But could we imagine these possibilities without the specious buttress of a transcendental structure (be that of God 1.0 or of Web 2.0)? To embrace again what Albert Camus called the splendour of this absurd world may allow us to recognize the absurdity of our desires for empowerment, and therefore allow us to differentiate between, on the one hand, the agonizingly slow, incremental steps which we may take (the overwhelmingly unrewarding efforts of actual participation we must make) towards the slim possibility of their eventual fulfilment, and, on the other, the spurious yet seductive claims by commercial, political and religious interests of easy routes through which (they declare) we are already being empowered. In essence, our only possibility of empowerment lies in our ability to distinguish between these two absurdities – the strenuous absurdity of individual empowerment as experienced, in fable at least, by Abraham and Jacob, and the absurd illusion of mediated empowerment as offered by Cowell, Wales and Zuckerberg – and then to embrace the former, comfortless state. This is not a passive acceptance of powerlessness but an active realization that material empowerment requires effortful commitment.

Having returned to Albert Camus, it is perhaps appropriate to conclude by recalling Camus's interpretation of the tale of a dead Greek king which concludes his great essay on absurdity, *The Myth of Sisyphus*. Camus finds, in the story of the man condemned for all eternity to roll a boulder to the top of a mountain, watch it roll back down again and then roll it up again, an apt enough analogy for the absurdity of the human condition. Yet Camus (1975: 111) discovers the possibility of nobility and joy in the Sisyphean situation: 'The struggle itself towards the heights is enough to fill a man's heart. We must imagine Sisyphus happy.'

The Sisyphean futility of the digital game's disempowering escapism or of the reality TV show's fifteen minutes of fame, or for that matter of the dilution of culture, society and democracy in the projects of Wikipedia, Facebook and electronic government, may most usefully be met, from Camus's perspective, by a recognition of that absurdity. This recognition may, eventually, be the extent of our potential for liberation. Or it may, after all, be only the beginning.

The acknowledgement that a certain degree of unhappiness may represent a perfectly sincere, rational, healthy and highly likely response to any particular set of social conditions (or indeed to the human condition *per se*) may offer an essential first step towards the recognition that it is not the only such response imaginable. It is only by accepting the near impossibility of happiness that such happiness may become, albeit improbably, possible.

Bibliography

Adams, D. (1979). *The Hitchhiker's Guide to the Galaxy*. New York: Harmony Books.

Adams, E. (2009). 'The philosophical roots of computer game design' in *Under the Mask: Perspectives on the Gamer*, University of Bedfordshire, 5 June 2009.

Adorno, T., and Horkheimer, M. (1979). *The Dialectic of Enlightenment* (trans. Cumming, J.). London: Verso.

Alas, J. (2007a). 'Card readers the only challenge in e-election' in *The Baltic Times*, 7 March 2007.

Alas, J. (2007b). 'Thumbs up for mobile voting' in *The Baltic Times*, 3 October 2007.

Allan, S. (2011). Keynote address in *Political Studies association Media and Politics Conference*, Bournemouth University, 3–4 November 2011.

Almond, M. (2011). 'How social networks trigger political downfall' in *The Sun*, 27 January 2011.

Althusser, L. (2006). *Lenin and Philosophy* (trans. Brewster, B.). Delhi: Aakar Books.

Anderson, J. (2011). *Wikipedia: The Company and its Founders*. Minnesota: ABDO Publishing.

Apter, T. (2010). 'Internet pornography' in *The Observer Magazine*, 10 April 2010.

Aris, B. (2004). 'Technological tiger' in *The Guardian*, 22 April 2004.

Aristotle (2004). *The Nicomachean Ethics* (trans. Thomson, J.). London: Penguin.

Arsenault, D., and Perron B. (2009). 'In the frame of the magic cycle: the circle(s) of gameplay' in *The Video Game Theory Reader 2* (ed. Perron, B., and Wolf, M.). London: Routledge. 109–131.

Ashton, J. (2012). 'Silicon supremo' in *The Independent*, 19 May 2012.

Åström, J. (2004). 'Digital democracy' in *Electronic Democracy* (ed. Gibson, R., Römmele, A., and Ward, S.). Abingdon: Routledge. 96–115.

Auden, W.H. (1979). *Selected Poems*. London: Faber & Faber.

Ayers, P., Matthews, C., and Yates, B. (2008). *How Wikipedia Works*. San Francisco: No Starch Press.

Baker, A. (2011). 'How Egypt's opposition got a more youthful mojo' in *Time*, 1 February 2011.

Baloun, K. (2006). *Inside Facebook: Life, Work and Visions of Greatness*. Self-published.

Bang, H. (2010). 'A new ruler meeting a new citizen: Culture governance and everyday making' in *Governance as Social and Political Communication* (ed. Bang, H.). Manchester: Manchester University Press. 241–266.

Barthes, R. (1974). *S/Z* (trans. Miller, R.). New York: Farrah.

Barthes, R. (1977). *Image-Music-Text* (trans. Heath, S.). London: Fontana.

Baudrillard, J. (1988). *America* (trans. Turner, C.). London: Verso.

Baudrillard, J. (1994). *Simulacra and Simulation* (trans. Glaser, S.). Michigan: University of Michigan Press.

Baudrillard, J. (1995). *The Gulf War Did Not Take Place* (trans. Patton, P.). Sydney: Power Publications.

Baudrillard, J. (2005). *The Intelligence of Evil or The Lucidity Pact* (trans. Turner, C.). Oxford: Berg.

Benjamin, W. (1992). *Illuminations* (trans. Zohn, H.). London: Fontana Press.

Benyon, J. (2012). 'England's urban disorder: The 2011 riots' in *Political Insight*, April 2012.

Bignell, J. (2005). *Big Brother: Reality TV in the Twenty-First Century*. Basingstoke: Palgrave Macmillan.

Biressi, A., and Nunn, H. (2005). *Reality TV: Realism and Revelation*. London: Wallflower Press.

Blanchot, M. (1971). *Friendship* (trans. Rottenberg, E.). Palo Alto, CA: Stanford University Press.

Blumler, J., and Gurevitch, M. (1995). *The Crisis in Public Communication*. London: Routledge.

Boellstorff, T. (2008). *Coming of Age in Second Life: An Anthropologist Explores the Virtually Human*. Princeton: Princeton University Press.

Bogost, I. (2006). *Unit Operations*. Cambridge, MA: MIT Press.

Bolter, J., and Grusin, R. (2000). *Remediation*. Cambridge, MA: MIT Press.

Boof, K. (2006). *Diary of a Lost Girl*. Aguanga, CA: Door of Kush Multimedia.

Borges, J. (1970). *Labyrinths* (trans. Yates, D., and Irby, J.). Harmondsworth: Penguin.

Boswell, J. (1986). *The Life of Samuel Johnson*. Harmondsworth: Penguin 1986.

Bourdieu, P. (1977). *Outline of a Theory of Practice* (trans. Nice, R.). Cambridge: Cambridge University Press.

Bourdieu, P. (1986). *Distinction: A Social Critique of the Judgement of Taste* (trans. Nice, R.). London: Routledge.

Bourdieu, P. (1991). *Language and Symbolic Power* (trans. Raymond, G., and Adamson, M.). Cambridge: Polity Press.

Bourdieu, P. (2005). 'The Mystery of ministry: From particular wills to the general will' (trans. Nice, R., and Wacquant, L.) in *Pierre Bourdieu and Democratic Politics* (ed. Wacquant, L.). Cambridge: Polity Press.

Boyd, C. (2004). 'Estonia opens politics to the web' in *BBC News Interactive*, 7 May 2004.

Boyd, D. (2006). 'Friends, Friendsters, and MySpace Top 8: Writing community into being on social network sites' in *First Monday* 11:12.

Boyd, D. (2008a). 'Facebook's privacy trainwreck: exposure, invasion and social convergence' in *Convergence* 14:1, 13–20.

Boyd, D. (2008b). *Taken Out of Context: American Teen Sociality in Networked Publics.* Doctoral dissertation, Berkeley, CA: University of California.

Boyd, D., and Heer, J. (2006). 'Profiles as conversation: Networked identity performance on Friendster' in *Proceedings of the Hawai'i International Conference on System Sciences.*

Brautigan, R. (1968). *The Pill versus the Springhill Mine Disaster.* New York: Delta Books.

Brecht, B. (1978). *Brecht on Theatre* (trans. Willett, J.). London: Methuen.

Breuer, F., and Trechsel, A. (2006). *Report for the Council of Europe: E-Voting in the 2005 local elections in Estonia.* Strasbourg: Council of Europe.

Broughton, J. (2008). *Wikipedia: The Missing Manual.* Sebastopol, CA: O'Reilly Media.

Brown, G. (2009). 'The internet is as vital as water and gas' in *The Times*, 16 June 2009.

Bruns, A. (2008). *Blogs, Wikipedia, Second Life, and Beyond: From Production to Produsage.* New York: Peter Lang.

Burke, M., Marlow, C., and Lento, T. (2010) 'Social network activity and social well-being' in *Proceedings of the 28th international conference on human factors in computing systems, Atlanta, Georgia.*

Burrell, I. (2010). 'It's not about how many pages. It's about how good they are' in *The Independent*, 20 December 2010.

Bynum, T., and Rogerson, S. (2004a). 'Ethics in the information age' in *Computer Ethics and Professional Responsibility* (ed. Bynum, T., and Rogerson, S.). Oxford: Blackwell Publishing. 1–13.

Bynum, T., and Rogerson, S. (2004b). 'Global information ethics' in *Computer Ethics and Professional Responsibility* (ed. Bynum, T., and Rogerson, S.). Oxford: Blackwell Publishing. 316–318.

Camus, A. (1975). *The Myth of Sisyphus* (trans. O'Brien, J.). Harmondsworth: Penguin.

Castronova, E. (2005). *Synthetic Worlds.* Chicago: University of Chicago Press.

Cellan-Jones, R. (2012). 'Is air leaking from the Facebook bubble?' in *BBC News Interactive*, 21 May 2012.

Chebib, N., and Sohail, R. (2011). 'The reasons social media contributed to the 2011 Egyptian revolution' in *International Journal of Business Research and Management*, 2:2, 139–162.

Cheng, L., Farnham, S., and Stone, L. (2002). 'Lessons learned: Building and deploying shared virtual environments' in *The Social Life of Avatars: Presence and Interaction in Shared Virtual Environments* (ed. Schroeder, R.). London: Springer. 90–111.

Chomsky, N. (1989). *Necessary Illusions.* London: Pluto Press.

Cicero, M. (2010). *Treatises on Friendship and Old Age* (trans. Shuckburgh, E.). Marston Gate: Hard Press.

Coleman, S., and Spiller, J. (2003). 'Exploring new media effects on representative democracy' in *The Journal of Legislative Studies* 9:3, 1–16.

Coleman, S. (2005a). 'E-democracy – what's the big idea?'. Manchester: British Council Governance Team.

Coleman, S. (2005b). 'Just how risky is online voting?' in *Information Polity* 10, 95–104.

Coleman, S. (2005c). 'The lonely citizen: Indirect representation in an age of networks' in *Political Communication* 22:2, 197–214.

Coleman, S. (2006). 'How the other half votes: *Big Brother* viewers and the 2005 general election' in *International Journal of Cultural Studies* 9:4, 457–479.

Coleman, S., and Blumler, J. (2009). *The Internet and Democratic Citizenship*. Cambridge: Cambridge University Press.

Collier, M. (2007). 'Estonia breaks into globalisation top 10' in *The Baltic Times*, 23 October 2007.

Condella, C. (2010). 'Why can't we be virtual friends?' in *Facebook and Philosophy* (ed. Wittkower, D.). Chicago: Carus Publishing. 111–122.

Couldry, N. (2010). 'Voice that matters' in *Media, Communication and Cultural Studies Association Conference*, London School of Economics, 6–8 January 2010.

Cummings, R. (2009). *Lazy Virtues: Teaching Writing in the Age of Wikipedia*. Nashville, TN: Vanderbilt University Press.

Curtice, J. (2009). Round table discussion in *Political Studies Association Media and Politics Group Conference*, University of Strathclyde, 5–6 November 2009.

Dalby, A. (2009). *The World and Wikipedia: How We Are Editing Reality*. Draycott: Siduri Books.

Dalton, S. (2011). 'The revolution will be streamed' in *The Times*, 25 March 2011.

Darley, A. (2000). *Visual Digital Culture*. London: Routledge.

Dawkins, R. (1989). *The Selfish Gene*. Oxford: Oxford University Press.

Day, E. (2010). 'Reality checks' in *The Observer Magazine*, 21 November 2010.

de Man, P. (1984). *The Rhetoric of Romanticism*. New York: Columbia University Press.

Derrida, J. (1981). *Dissemination* (trans. Johnson, B.). London: Athlone Press.

Derrida, J. (1987). Interview in *Criticism and Society* (ed. Salusinszky, I.). London: Methuen.

Derrida, J. (1989). *Memoires for Paul de Man* (trans. Kamuf, P.). New York: Columbia University Press.

Derrida, J. (2005). *The Politics of Friendship* (trans. Collins, G.) London: Verso.

Dickens, C. (1997). *David Copperfield*. London: Penguin.

Dixon, S. (2007). *Digital Performance: A History of New Media in Theatre, Dance Performance Art and Installation*. Cambridge, MA: MIT Press.

Downing, J. (2012). Keynote address in *Media, Communication and Cultural Studies Association Conference*, University of Bedfordshire, 11–13 January 2012.

Doyle, W., and Fraser, M. (2010). 'Facebook, surveillance and power' in *Facebook and Philosophy* (ed. Wittkower, D.). Chicago: Carus Publishing. 215–230.

Dunbar, R. (2010). 'How many friends does one person need?', paper presented to the Royal Society for the Encouragement of the Arts, Manufactures and Commerce, London, 18 February 2010.

Eco, U. (1995), 'Ur-Fascism' in *The New York Review of Books*, 22 June 1995.

Eliot, T.S. (1925). 'The Hollow Men' in *Poems 1909–1925*. London: Faber & Faber.

Ellison, N., Steinfeld, C., and Lampe, C. (2007). 'The benefits of Facebook friends: social capital and college students' use of online social network sites' in *Journal of Mediated Communication* 12, 1143–1168.

Ernsdorff, M., and Berbec, A. (2007). 'Estonia: The short road to e-government and e-democracy', in *E-government in Europe* (ed. Nixon, P., and Koutrakou, V.). Abingdon: Routledge. 171–183.

Estonian National Electoral Committee (2007). *Parliamentary elections 2007: Statistics of e-voting*. Tallinn: Estonian National Electoral Committee.

European Commission (2005). *Online government is now a reality almost everywhere in the EU*. Brussels: European Commission, 8 March 2005.

Fanon, F. (1990), *The Wretched of the Earth* (trans. Farrington, C.). Harmondsworth: Penguin.

Fisk, R. (2011), 'Egypt's day of reckoning' in *The Independent*, 28 January 2011.

Fiske, J. (1987). *Television Culture*. London: Routledge.

Fletcher, M. (2011). 'Crowds salute the Facebook dreamer who led his nation' in *The Times*, 9 February 2011.

Fossato, F., and Lloyd, J., with Verkhovsky, A. (2008). *The Web that Failed*. Oxford: Reuters Institute for the Study of Journalism.

Foucault, M. (1991). *Discipline and Punish* (trans. Sheridan, A.). Harmondsworth: Penguin.

Foucault, M. (1998). *The Will to Knowledge* (trans. Hurley, R.). London: Penguin.

Frasca, G. (2007). *Play the Message: Play, Game and Videogame Rhetoric*. PhD dissertation, Copenhagen: IT University of Copenhagen.

Frau-Meigs, D. (2007). 'Convergence, Internet governance and cultural diversity' in *Ambivalence towards Convergence* (ed. Storsul, T., and Stuedahl, D.). Göteborg: Nordicom. 33–53.

Freud, S. (1984). *On Metapsychology* (trans. Strachey, J.). Harmondsworth: Penguin.

Freud, S. (1985). *Art and Literature* (trans. Strachey, J.). Harmondsworth: Penguin.

Fuchs, C. (2008). *Internet and Society: Social Theory in the Information Age*. New York: Routledge.

Fuchs, C. (2011). 'New media, Web 2.0 and surveillance' in *Sociology Compass* 5:2, 134–147.

Fuller, T. (2004). 'In Estonia: E-banking, e-commerce, e-government' in *International Herald Tribune*, 13 September 2004.

Gaber, I. (2010). 'Political journalism in retreat' in *Political Studies Association Conference*, Edinburgh, 29 March – 1 April 2010.

Genvo, S. (2009). 'Understanding digital playability' in *The Video Game Theory Reader 2* (ed. Perron, B., and Wolf, M.). London: Routledge. 133–149.

Gerges, F. (2011). 'Cry freedom' in *The Mirror*, 21 February 2011.

Ghannam, J. (2011). *Social Media in the Arab World*. Washington, D.C.: Center for International Media Assistance.

Ghonim, W. (2012). *Revolution 2.0*. New York: Houghton Mifflin Harcourt Publishing.

Gibson, R., Lusoli, W., Römmele, A., and Ward, S. (2004). 'Representative democracy and the Internet' in *Electronic Democracy* (ed. Gibson, R., Römmele, A., and Ward, S.). Abingdon: Routledge. 1–16.

Giles, J. (2005). 'Internet encyclopedias go head to head' in *Nature* 438, 900–901.

Glassman, M., Straus, J., and Shogan, C. (2010). *Social Networking and Constituent Communications*. Washington, D.C.: Congressional Research Service.

Gove, M. (2001). 'America awakes to terrorism by timetable – and the darkest national catastrophe' in *The Times*, 12 September 2001.

Greenstein, F. (1967). 'The impact of personality on politics' in *The American Political Science Review* 61:3, 629–641.

Gregg, M. (2011). *Work's Intimacy*. Cambridge: Polity Press.

Gur, O. (2010). 'Comparing social network sites and past social systems' in *Media, Communication and Cultural Studies Association Conference*, London School of Economics, 6–8 January 2010.

Habermas, J. (1989). *The Structural Transformation of the Public Sphere* (trans. Burger, T.). Cambridge: Polity Press.

Hall, J. (2001). *Online Journalism*. London: Pluto Press.

Hall, S. (1980). 'Encoding/decoding' in *Culture, Media, Language* (ed. Hall, S., Hobson, D., Lowe, A., and Willis, P.). London: Hutchinson. 128–139.

Hand, M. (2008). *Making Digital Cultures: Access, Interactivity, and Authenticity*. Aldershot: Ashgate Publishing.

Hands, J. (2011). *@ is for Activism*. London: Pluto Press.

Harding, A. (2012). 'Joseph Kony campaign under fire' in *BBC News Interactive*, 8 March 2012.

Hayles, N. (2000). 'The condition of virtuality' in *The Digital Dialectic* (ed. Lunenfeld, P.). Cambridge, MA: MIT Press. 68–94.

Hayward, D. (2008). 'Games and culture' in *Under the Mask: Perspectives on the Gamer*, University of Bedfordshire, 7 June 2008.

Herman, E., and Chomsky, N. (2002). *Manufacturing Consent: The Political Economy of the Mass Media*. New York: Pantheon.

Hill, A. (2005). *Reality TV: Audiences and Popular Factual Television*. London: Routledge.

Hodgkinson, T. (2008). 'With friends like these ...' in *The Guardian*, 14 January 2008.

Hodson, R., and Sullivan, T. (2012). *The Social Organization of Work*. Belmont, CA: Wadsworth.

Hoggard, L. (2010). 'Tough on the causes of cat crime' in *The Independent*, 26 August 2010.

Hughes, R. (1991). *The Shock of the New*. London: BBC Books.

Hunicke, R. (2008). 'The modern age of gaming' in *Lift08*, Geneva, 6–8 February 2008.

Hyde, M. (2010). 'Cameron's plan to plunder the oeuvre of Simon Cowell' in *The Guardian*, 1 January 2010.

Jackson, M. (2011). 'Facebook: five things to avoid' in *BBC News Interactive*, 16 June 2011.

Jackson, P. (2011). 'England riots: Dangers behind false rumours' in *BBC News Interactive*, 12 August 2011.

Jameson, F. (1991). *Postmodernism, or, The Cultural Logic of Late Capitalism*. London: Verso.

Jameson, F. (2005). *Archaeologies of the Future*. London: Verso.

Jeffries, S. (2011). 'Sock puppets, twitterjacking and the art of digital fakery' in *The Guardian*, 29 September 2011.

Johnson, S. (1755). *A Dictionary of the English Language*. London: Knapton, Longman, Hitch, Hawes, Millar and Dodsley.

Johnson, S. (1951). *The Selected Letters of Samuel Johnson*. Oxford: Oxford University Press.

Johnson, S. (1971). *Rasselas, Poems, and Selected Prose*. San Francisco: Rinehart Press.

Jones, J., and Salter, L. (2012). *Digital Journalism*. London: Sage.

Kanai, R., Bahrami, B., Roylance, R., and Rees, G. (2011). 'Online social network size is reflected in human brain structure' in *Proceedings of the Royal Society B (Biological Sciences)*, October 2011.

Kaur, N. (2007). *Big Brother: The Inside Story*. London: Virgin Books.

Keen, A. (2008). *The Cult of the Amateur*. London: Nicholas Brealey Publishing.

Kelsey, T. (2010). *Social Networking Spaces*. New York: Springer.

Khamis, S., and Vaughn, K. (2011). 'Cyberactivism in the Egyptian revolution: How civic engagement and citizen journalism tilted the balance' in *Arab Media and Society* 14.

Khoury, D. (2011) 'Social media and the revolutions: How the Internet revived the Arab public sphere and digitalized activism' in *Perspectives* 2, 80–86.

Kierkegaard, S. (2006). *Fear and Trembling* (trans. Walsh, S.). Cambridge: Cambridge University Press.

Kierkegaard, S. (2009). *Concluding unscientific postscript* (trans. Hannay, A.). Cambridge: Cambridge University Press.

Kilborn, R. (2003). *Staging the Real*. Manchester: Manchester University Press.

King, G. (2005). *The Spectacle of the Real*. Bristol: Intellect.

King, G., and Krzywinska, T. (2006). *Tomb Raiders and Space Invaders*. London: I.B. Tauris.

Kirkpatrick, D. (2011). *The Facebook Effect*. London: Random House.

Klimmt, C., and Hartmann, T. (2006). 'Effectance, self-efficacy and the motivation to play video games' in *Playing Video Games: Motives, Responses and Consequences* (ed. Vorderer, P., and Bryant, J.). Mahwah, NJ: Lawrence Erlbaum. 133–145.

Konzack, L. (2009). 'Philosophical game design' in *The Video Game Theory Reader 2* (ed. Perron, B., and Wolf, M.). London: Routledge. 33–44.

Koskinen, I. (2007). 'The design professions in convergence' in *Ambivalence towards Convergence* (ed. Storsul, T., and Stuedahl, D.). Göteborg: Nordicom. 117–128.

Krzywinska, T. (2008). 'Reanimating H.P. Lovecraft' in *Under the Mask: Perspectives on the Gamer*, University of Bedfordshire, 7 June 2008.

Lang, D., Böhm, C., and Naumann, F. (2010). *Extracting Structured Information from Wikipedia Articles to Populate Infoboxes*. Potsdam: University of Potsdam.

Leaning, M. (2011). 'Understanding blogs: just another media form' in *The End of Journalism* (ed. Charles, A., and Stewart, G.). Oxford: Peter Lang. 87–102.

LeBor, A. (2011). 'Despots beware' in *The Times*, 17 January 2011.

Levi, P. (1989). *The Drowned and the Saved* (trans. Rosenthal, R.). London: Abacus.

Li, Z. (2004). *The Potential of America's Army the Video Game as Civilian-Military Public Sphere*. MSc dissertation, Cambridge: Massachusetts Institute of Technology.

Lifton, R., and Markusen, E. (1991). *The Genocidal Mentality*. London: Macmillan.

Macleod, H. (2011). 'On the frontline with Syria's switched-on cyber activists' in *The Guardian*, 16 April 2011.

Madise, Ü., Vinkel, P., and Maaten, E. (2006). *Internet Voting at the Elections of Local Government Councils on October 2005*. Tallinn: Estonian National Electoral Committee.

Madise, Ü. (2007). *Internet Voting in Estonia Free and Fair Elections*. Tallinn: Estonian National Electoral Committee.

Maitles, H., and Gilchrist, I. (2005). 'We're citizens now!: The development of positive values through a democratic approach to learning' in *Journal for Critical Education Policy Studies* 3:1.

Manovich, L. (2001). *The Language of New Media*. Cambridge, MA: MIT Press.

Margolis, M. (2007). 'E-government and democratic politics', in *E-government in Europe* (ed. Nixon, P., and Koutrakou, V.). Abingdon: Routledge. 1–18.

Martens, T. (2007). *Internet Voting in Practice*. Tallinn: Estonian National Electoral Committee.

May, P. (2010). *Virtually Dead*. Scottsdale: Poisoned Pen Press.

McGonigal, J. (2008). 'Reality is broken' in *Game Developer's Conference*, San Francisco, 22 February 2008.

McLuhan, M. (2001). *Understanding Media*. London: Routledge.

McQueen, D. (1998). *Television*. London: Arnold.

Middleton, C. (1999). 'Ethics man' in *Business and Technology*, January 1999. 22–27.

Mikecz, R. (2005). 'Ambition versus pragmatism' in *EU Enlargement – One Year On* (ed. Charles, A.). Tallinn: Audentes University. 146–156.

Miller, D. (2011). *Tales from Facebook*. Cambridge: Polity Press.

Montaigne, M. de (1991). *The Complete Essays* (trans. Screech, M.). Harmondsworth: Penguin.

Morozov, E. (2011a). *The Net Delusion*. London: Allen Lane.

Morozov, E. (2011b). 'Facebook and Twitter are just places revolutionaries go' in *The Guardian*, 7 March 2011.

Moylan, T. (1986). *Demand the Impossible: Science Fiction and the Utopian Imagination*. London: Methuen.

Murphy, S. (2009). 'This is intelligent television: Early video games and television in the emergence of the personal computer' in *The Video Game Theory Reader 2* (ed. Perron, B., and Wolf, M.). London: Routledge. 197–212.

Myers, D. (2009). 'The video game aesthetic: play as form' in *The Video Game Theory Reader 2* (ed. Perron, B., and Wolf, M.). London: Routledge. 45–63.

Nature (2006). 'Britannica attacks' in *Nature* 440, 582.

Needham, C. (2004). 'The citizen as consumer: E-government in the United Kingdom and the United States' in *Electronic Democracy* (ed. Gibson, R., Römmele, A., and Ward. S.). Abingdon: Routledge. 43–69.

Neumann, P. (2004). 'Computer security and human values' in *Computer Ethics and Professional Responsibility* (ed. Bynum, T., and Rogerson, S.). Oxford: Blackwell Publishing. 208–226.

Newman, J. (2002). 'The myth of the ergodic videogame' in *Game Studies* 2:1.

Nieborg, D. (2006). 'First person paradoxes – the logic of war in computer games' in *Game Set and Match II: On Computer Games, Advanced Geometries and Digital Technologies* (ed. Oosterhuis, K., and Feireiss, L.). Rotterdam: Episode Publishers. 107–115.

Nietzsche, F. (1969). *Thus Spoke Zarathustra* (trans. Hollingdale, R.). Harmondsworth: Penguin.

Nietzsche, F. (2008). *Human, All Too Human* (trans. Zimmern, H.). Ware: Words-
worth Editions.

Nixon, P. (2007). 'Ctrl, alt, delete: Rebooting the European Union via e-government'
in *E-government in Europe* (ed. Nixon, P., and Koutrakou, V.). Abingdon: Routledge. 19–32.

Nixon, P., and Koutrakou, V. (2007). 'Introduction' in *E-government in Europe* (ed.
Nixon, P., and Koutrakou, V.). Abingdon: Routledge. xviii–xxviii.

Oberholzer-Gee, F., and Waldfogel, J. (2005). 'Strength in numbers: Group size and
political mobilisation' in *Journal of Law and Economics* 158, 73–91.

O'Loughlin, E. (2012). 'Revolution 2.0' in *The Daily Telegraph*, 13 January 2012.

Orwell, G. (1951). *Animal Farm*. Harmondsworth: Penguin.

Osbourne, A. (2011). 'Moscow airport bombing: Why a terrorist mastermind is sending chills down spines' in *Daily Telegraph*, 29 January 2011.

OSCE (2007). *Republic of Estonia Parliamentary Elections 4 March 2007*. Warsaw:
OSCE / Office for Democratic Institutions and Human Rights.

OSCE (2011). *Estonia Parliamentary Elections 6 March 2011*. Warsaw: OSCE / Office
for Democratic Institutions and Human Rights.

O'Sullivan, D. (2009). *Wikipedia: A New Community of Practice?*. Farnham: Ashgate
Publishing.

Ouellette, L. (2009). '"Take responsibility for yourself": Judge Judy and the neoliberal
citizen' in *Reality TV: Remaking Television Culture* (ed. Murray, S., and Ouellette, L.). New York: New York University Press. 223–242.

Palfrey, J., and Gasser, U. (2008). *Born Digital*. New York: Basic Books.

Papacharissi, Z. (2010). *A Private Sphere: Democracy in a Digital Age*. Cambridge:
Polity Press.

Pleace, N. (2007). 'E-government and the United Kingdom' in *E-government in Europe*
(ed. Nixon, P., and Koutrakou, V.). Abingdon: Routledge. 61–74.

Poole, E. (2002). *Reporting Islam*. London: I.B. Tauris.

Pratchett, L. (2007). 'Local e-democracy in Europe: Democratic x-ray as the basis for
comparative analysis' in *International Conference on Direct Democracy in Latin
America*, Buenos Aires, 14–15 March 2007.

Price, J.H. (2010). 'The new media revolution in Egypt: |Understanding the failures
of the past and looking towards the possibilities of the future' in *Democracy &
Society* 7:12. 1, 18–20.

Price, S. (2010). *Brute Reality*. London: Pluto Press.

Putnam, R. (2000). *Bowling Alone*. New York: Simon & Schuster.

Raab, C., and Bellamy, C. (2004). 'Electronic democracy and the "mixed polity"',
in (eds), *Electronic Democracy* (ed. Gibson, R., Römmele, A., and Ward, S.).
Abingdon: Routledge. 17–42.

Radsch, C. (2011). 'Blogosphere and social media' in *Seismic Shift: Understanding Change in the Middle East* (ed. Laipson, E.). Washington, DC: The Henry L. Stimson Center. 67–81.

Reagle, J. (2010). *Good Faith Collaboration: The Culture of Wikipedia*. Cambridge, MA: MIT Press.

Redden, J., and Witschge, T. (2010). 'A new news order?' in *New Media, Old News* (ed. Fenton, N.). London: Sage. 171–186.

Rice, J. (2009). *The Church of Facebook: How the Hyperconnected are Redefining Community*. Colorado Springs: David C. Cook.

Roberts, G. (2006). 'Orange suspends blogger over his "Lefty Lexicon"' in *The Independent*, 18 August 2006.

Robinson, N. (2006). 'Prescott for dummies' in *BBC News Interactive*, 5 July 2006.

Sartre, J.-P. (1969). *Being and Nothingness* (trans. Barnes, H.). London: Methuen.

Sartre, J.-P. (2000). *Nausea* (trans. Baldick, R.). London: Penguin.

Savigny, H., and Temple, M. (2010). 'Politics marketing models: The curious incident of the dog that doesn't bark' in *Political Studies* 58:5, 1049–1064.

Schmeink, L. (2008). 'The horror of being human: Dystopia, alternate realities and post-human societies in recent videogames' in *Bridges to Utopia*, University of Limerick, 3–5 July 2008.

Schmemann, S. (2001). 'Hijacked jets destroy twin towers and hit Pentagon' in *The New York Times*, 12 September 2001.

Sherwin, A. (2012). 'Andrew Lloyd-Webber blames Eurovision failures on "racist" Eastern Europe' in *The Independent*, 3 July 2012.

Siibak, A. (2009). 'Constructing the self through the photo selection – visual impression management on social networking websites' in *Cyberpsychology: Journal of Psychosocial Research on Cyberspace*, 3:1.

Silverman, J. (2003). 'YouTube if you want to: New media, investigative journalism and social control' in *The End of Journalism* (ed. Charles, A., and Stewart, G.). Oxford: Peter Lang. 51–61.

Silverman, K. (1984). *The Subject of Semiotics*. Oxford: Oxford University Press.

Singh, A. (2011). 'Role of Twitter and Facebook in Arab Spring uprising overstated, says Hisham Matar' in *Daily Telegraph*, 11 July 2011.

Smallwood, J. (2010). 'Facebook: should parents "friend" their children?' in *BBC News Interactive*, 10 December 2010.

Soon, C., Brass, M., Heinze H.-J., and Haynes, J.-D. (2008). 'Unconscious determinants of free decisions in the human brain' in *Nature Neuroscience* 11, 543–545.

Stewart, G. (2011). '"I cant belive a war started and Wikipedia sleeps": making news with an online encyclopaedia' in *The End of Journalism* (ed. Charles, A., and Stewart, G.). Oxford: Peter Lang. 139–158.

Storsul, T., and Stuedahl, D. (2007). 'Introduction' in *Ambivalence towards Convergence* (ed. Storsul, T., and Stuedahl, D.). Göteborg: Nordicom. 9–16.

Swanson, D., and Mancini, P. (1996). 'Patterns of modern electoral campaigning and their consequences' in *Politics, Media and Modern Democracy* (ed. Swanson, D., and Mancini, P.). Westport: Praeger. 247–276.

Talbot, M. (2007). *Media Discourse: Representation and Interaction.* Edinburgh: Edinburgh University Press.

Taylor, J. (2010). 'Kanye West follows only one – but who is Steven of Coventry?' in *The Independent*, 2 August 2010.

Tedesco, M. (2010). 'The Friendship that makes no demands' in *Facebook and Philosophy* (ed. Wittkower, D.). Chicago: Carus Publishing. 123–134.

Tiffin, J., and Rajasingham, L. (1995). *In Search of the Virtual Class.* Abingdon: Routledge.

Tiffin, J., and Rajasingham, L. (2003). *The Global Virtual University.* London: RoutledgeFalmer.

Tincknell, E., and Raghuram, P. (2004). '*Big Brother*: reconfiguring the "active" audience of cultural studies?' in *Understanding Reality Television* (ed. Holmes, S., and Jermyn, D.). London: Routledge. 252–269.

Trechsel, A. (2007). *Report for the Council of Europe: Internet voting in the March 2007 Parliamentary Elections in Estonia.* Strasbourg: Council of Europe.

Trost, C., and Grossmann, M. (2005). *Win the Right Way.* Berkeley: Public Policy Press.

Tubalkain-Trell, M. (2008). 'Report calls out underachieving minister' in *The Baltic Times*, 9 July 2008.

Tumber, H., and Webster, F. (2006). *Journalists Under Fire.* London: Sage.

UNHCR (2009). *Freedom on the Net 2009 – Estonia.* Washington, D.C.: Freedom House.

UN World Public Sector Report (2003). *E-Government at the Crossroads.* New York: United Nations.

Vaiga, L. (2008). 'Lithuania says no to e-voting' in *The Baltic Times*, 23 January 2008.

Valenzuela, S., Park, N., and Kee, K. (2009). 'Is there social capital in a social network site?' in *Journal of Computer-Mediated Communication* 14, 875–901.

van Ham, P. (2001). 'The rise of the brand state: The postmodern politics of image and reputation' in *Foreign Affairs* 80:5, 2–6.

van Zoonen, L. (2004). 'Desire and resistance: *Big Brother* in the Dutch public sphere' in *Big Brother International: Formats, Critics and Publics* (ed. Mathijs, E., and Jones, J.). London: Wallflower Press. 16–24.

van Zoonen, L. (2010). 'Islam on the popular battlefield: Performing politics and religion on YouTube' in *Political Studies Association Media and Politics Group Conference*, Loughborough University, 4–5 November 2010.

Vasaloua, A., Joinson, A., Bänziger, T., Goldie, P., and Pitt, J. (2008). 'Avatars in social media: Balancing accuracy, playfulness and embodied messages' in *International Journal of Human-Computer-Studies* 66:11, 801–811.

Watt, N. (2012). 'Rupert Murdoch pressured Tony Blair over Iraq, says Alastair Campbell' in *The Guardian*, 15 June 2012.

Webb, B. (2012). 'Should Louise Mensch be setting up a Twitter rival on the side?' in *The Guardian*, 20 June 2012.

Weissmann, G. (2009). 'Walter Benjamin and Biz Stone: The scientific paper in the age of Twitter' in *The Journal of the Federation of American Societies for Experimental Biology* 23, 2015–2018.

Wesch, M. (2008). 'The Machine is Us/ing Us', paper presented at the US. Library of Congress, 23 June 2008.

Wilkinson, H. (2010). 'Spin is dead! Long live spin' in *Political Insight* 1:2, 45–47.

Williams, D. (2011). 'Rioters call for Gaddafi to go' in *Daily Mail*, 17 February 2011.

Williams, Z. (2003). 'The final irony' in *The Guardian*, 28 June 2003.

Williamson, A. (2009). *MPs Online: Connecting with Constituents*. London: Hansard Society.

Wilson, G. (2011). 'I predict a rioter' in *The Sun*, 25 October 2011.

Wright, S. (2011). 'Downing Street e-petitions: enhancing political communication and democracy?' in *Political Studies Association Conference*, London, 19–21 April 2011.

Wright, T., Boria, E., and Breidenbach, P. (2002). 'Creative player actions in FPS online video games: Playing Counter-Strike' in *Game Studies* 2:2.

Xenakis, A., and Macintosh, A. (2007). 'A methodology for the redesign of the electoral process to an e-electoral process' in *International Journal Electronic Governance* 1:1, 4–16.

Žižek, S. (1989). *The Sublime Object of Ideology*. London: Verso Books.

Žižek, S. (1999). 'The Cyberspace Real' in *Problemi* 37:3/4, 5–16.

Žižek, S. (2002). *Welcome to the Desert of the Real!*. London: Verso Books.

Žižek, S. (2008). *Violence*. London: Profile Books.

Zywica, J., and Danowski, J. (2008). 'The faces of Facebookers: Investigating social enhancement and social compensation hypotheses' in *Journal of Computer-Mediated Communication* 14, 1–34.

Index